# Airman's Odyssey

*A Trilogy Comprising*

WIND, SAND AND STARS

NIGHT FLIGHT

FLIGHT TO ARRAS

# AIRMAN'S ODYSSEY

by

Antoine de Saint-Exupéry

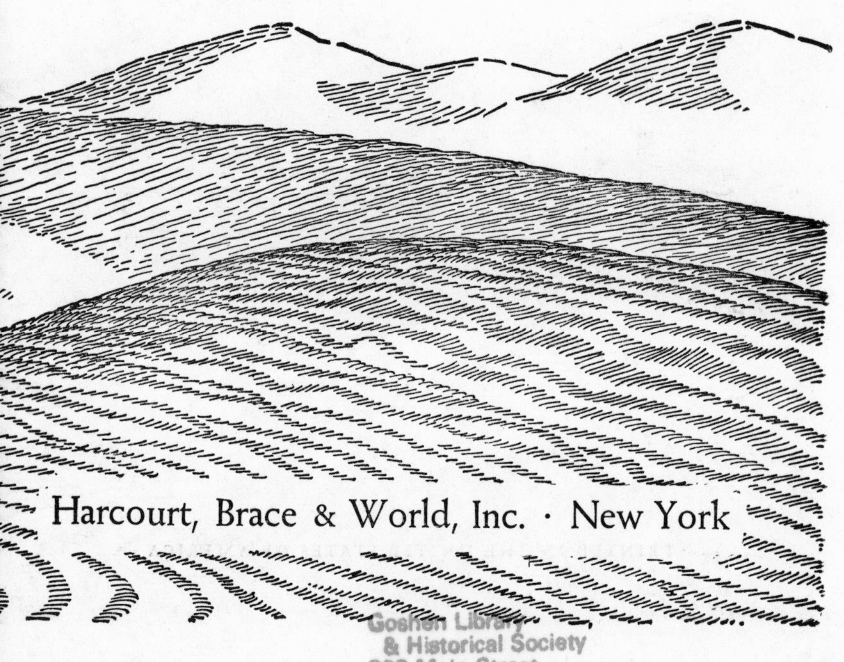

Harcourt, Brace & World, Inc. · New York

*Tenth Printing,*

PRINTED IN THE UNITED STATES OF AMERICA

*Table of Contents*

# WIND, SAND AND STARS

*Translated from the French*
*By Lewis Galantière*

# I

## *The Craft*

In 1926 I was enrolled as student airline pilot by the Latécoère Company, the predecessors of Aéropostale (now Air France) in the operation of the line between Toulouse, in southwestern France, and Dakar, in French West Africa. I was learning the craft, undergoing an apprenticeship served by all young pilots before they were allowed to carry the mails. We took ships up on trial spins, made meek little hops between Toulouse and Perpignan, and had dreary lessons in meteorology in a freezing hangar. We lived in fear of the mountains of Spain, over which we had yet to fly, and in awe of our elders.

These veterans were to be seen in the field restaurant—gruff, not particularly approachable, and inclined somewhat to condescension when giving us the benefit of their experience. When one of them landed, rain-soaked and behind schedule, from Alicante or Casablanca, and one of us asked humble questions about his flight, the very curtness of his replies on these tempestuous days was matter enough out of which to build a fabulous world filled with snares and pitfalls, with cliffs suddenly looming out of fog and whirling air-currents of a strength to uproot cedars. Black dragons guarded the mouths of the valleys and clusters of lightning crowned the crests—for our elders were always at some pains to feed our reverence. But from time to time one or another of them, eternally to be revered, would fail to come back.

I remember, once, a homecoming of Bury, he who was later to

die in a spur of the Pyrenees. He came into the restaurant, sat down at the common table, and went stolidly at his food, shoulders still bowed by the fatigue of his recent trial. It was at the end of one of those foul days when from end to end of the line the skies are filled with dirty weather, when the mountains seem to a pilot to be wallowing in slime like exploded cannon on the decks of an antique man-o'-war.

I stared at Bury, swallowed my saliva, and ventured after a bit to ask if he had had a hard flight. Bury, bent over his plate in frowning absorption, could not hear me. In those days we flew open ships and thrust our heads out round the windshield, in bad weather, to take our bearings: the wind that whistled in our ears was a long time clearing out of our heads. Finally Bury looked up, seemed to understand me, to think back to what I was referring to, and suddenly he gave a bright laugh. This brief burst of laughter, from a man who laughed little, startled me. For a moment his weary being was bright with it. But he spoke no word, lowered his head, and went on chewing in silence. And in that dismal restaurant, surrounded by the simple government clerks who sat there repairing the wear and tear of their humble daily tasks, my broad-shouldered messmate seemed to me strangely noble; beneath his rough hide I could discern the angel who had vanquished the dragon.

The night came when it was my turn to be called to the field manager's room.

He said: "You leave tomorrow."

I stood motionless, waiting for him to dismiss me. After a moment of silence he added:

"I take it you know the regulations?"

In those days the motor was not what it is today. It would drop out, for example, without warning and with a great rattle like the crash of crockery. And one would simply throw in one's hand: there was no hope of refuge on the rocky crust of Spain. "Here," we used to say, "when your motor goes, your ship goes, too."

An airplane, of course, can be replaced. Still, the important thing was to avoid a collision with the range; and blind flying through a sea of clouds in the mountain zones was subject to the severest penalties. A pilot in trouble who buried himself in the white cotton-wool of the clouds might all unseeing run straight into a peak. This was why, that night, the deliberate voice repeated insistently its warning:

"Navigating by the compass in a sea of clouds over Spain is all very well, it is very dashing, but—"

And I was struck by the graphic image:

"But you want to remember that below the sea of clouds lies eternity."

And suddenly that tranquil cloud-world, that world so harmless and simple that one sees below on rising out of the clouds, took on in my eyes a new quality. That peaceful world became a pitfall. I imagined the immense white pitfall spread beneath me. Below it reigned not what one might think—not the agitation of men, not the living tumult and bustle of cities, but a silence even more absolute than in the clouds, a peace even more final. This viscous whiteness became in my mind the frontier between the real and the unreal, between the known and the unknowable. Already I was beginning to realize that a spectacle has no meaning except it be seen through the glass of a culture, a civilization, a craft. Mountaineers too know the sea of clouds, yet it does not seem to them the fabulous curtain it is to me.

When I left that room I was filled with a childish pride. Now it was my turn to take on at dawn the responsibility of a cargo of passengers and the African mails. But at the same time I felt very meek. I felt myself ill-prepared for this responsibility. Spain was poor in emergency fields; we had no radio; and I was troubled lest when I got into difficulty I should not know where to hunt a landing-place. Staring at the aridity of my maps, I could see no help in them; and so, with a heart full of shyness and pride, I fled to spend this night of vigil with my friend Guillaumet. Guillaumet had been over the route before me. He knew all the dodges by which one got hold of the keys to Spain. I should have to be initiated by Guillaumet.

When I walked in he looked up and smiled.

"I know all about it," he said. "How do you feel?"

He went to a cupboard and came back with glasses and a bottle of port, still smiling.

"We'll drink to it. Don't worry. It's easier than you think."

Guillaumet exuded confidence the way a lamp gives off light. He was himself later on to break the record for postal crossings in the Andes and the South Atlantic. On this night, sitting in his shirtsleeves, his arms folded in the lamplight, smiling the most heartening of smiles, he said to me simply:

"You'll be bothered from time to time by storms, fog, snow. When you are, think of those who went through it before you, and say to yourself, 'What they could do, I can do.' "

I spread out my maps and asked him hesitantly if he would mind going over the hop with me. And there, bent over in the lamplight, shoulder to shoulder with the veteran, I felt a sort of schoolboy peace.

But what a strange lesson in geography I was given! Guillaumet did not teach Spain to me, he made the country my friend. He did not talk about provinces, or peoples, or livestock. Instead of telling me about Guadix, he spoke of three orange-trees on the edge of the town: "Beware of those trees. Better mark them on the map." And those three orange-trees seemed to me thenceforth higher than the Sierra Nevada.

He did not talk about Lorca, but about a humble farm near Lorca, a living farm with its farmer and the farmer's wife. And this tiny, this remote couple, living a thousand miles from where we sat, took on a universal importance. Settled on the slope of a mountain, they watched like lighthouse-keepers beneath the stars, ever on the lookout to succor men.

The details that we drew up from oblivion, from their inconceivable remoteness, no geographer had been concerned to explore. Because it washed the banks of great cities, the Ebro River was of interest to map-makers. But what had they to do with that brook running secretly through the water-weeds to the west of Motril, that brook nourishing a mere score or two of flowers?

"Careful of that brook: it breaks up the whole field. Mark it on your map." Ah, I was to remember that serpent in the grass near Motril! It looked like nothing at all, and its faint murmur sang to no more than a few frogs; but it slept with one eye open. Stretching its length along the grasses in the paradise of that emergency landing-field, it lay in wait for me a thousand miles from where I sat. Given the chance, it would transform me into a flaming candelabra. And those thirty valorous sheep ready to charge me on the slope of a hill! Now that I knew about them I could brace myself to meet them.

"You think the meadow empty, and suddenly bang! there are thirty sheep in your wheels." An astounded smile was all I could summon in the face of so cruel a threat.

Little by little, under the lamp, the Spain of my map became a sort of fairyland. The crosses I marked to indicate safety zones and traps were so many buoys and beacons. I charted the farmer, the thirty sheep, the brook. And, exactly where she stood, I set a buoy to mark the shepherdess forgotten by the geographers.

When I left Guillaumet on that freezing winter night, I felt the need of a brisk walk. I turned up my coat collar, and as I strode among the indifferent passers-by I was escorting a fervor as tender as if I had just fallen in love. To be brushing past these strangers with that marvelous secret in my heart filled me with pride. I seemed to myself a sentinel standing guard over a sleeping camp. These passers-by knew nothing about me, yet it was to me that, in their mail pouches, they were about to confide the weightiest cares of their hearts and their trade. Into my hands were they about to entrust their hopes. And I, muffled up in my cloak, walked among them like a shepherd, though they were unaware of my solicitude.

Nor were they receiving any of those messages now being despatched to me by the night. For this snowstorm that was gathering, and that was to burden my first flight, concerned my frail flesh, not theirs. What could they know of those stars that one by one were going out? I alone was in the confidence of the stars. To me alone news was being sent of the enemy's position before the hour of battle. My footfall rang in a universe that was not theirs.

These messages of such grave concern were reaching me as I walked between rows of lighted shop-windows, and those windows on that night seemed a display of all that was good on earth, of a paradise of sweet things. In the sight of all this happiness, I tasted the proud intoxication of renunciation. I was a warrior in danger. What meaning could they have for me, these flashing crystals meant for men's festivities, these lamps whose glow was to shelter men's meditations, these cozy furs out of which were to emerge pathetically beautiful solicitous faces? I was still wrapped in the aura of friendship, dazed a little like a child on Christmas Eve, expectant of surprise and palpitatingly prepared for happiness; and yet already I was soaked in spray; a mail pilot, I was already nibbling the bitter pulp of night flight.

It was three in the morning when they woke me. I thrust the shutters open with a dry snap, saw that rain was falling on the town, and got soberly into my harness. A half-hour later I was out on the pavement shining with rain, sitting on my little valise and waiting for the bus that was to pick me up. So many other flyers before me, on their day of ordination, had undergone this humble wait with beating heart.

Finally I saw the old-fashioned vehicle come round the corner and heard its tinny rattle. Like those who had gone before me, I squeezed in between a sleepy customs guard and a few glum government clerks. The bus smelled musty, smelled of the dust of government offices into which the life of a man sinks as into a quicksand. It stopped every five hundred yards to take on another scrivener, another guard, another inspector.

Those in the bus who had already gone back to sleep responded with a vague grunt to the greeting of the newcomer, while he crowded in as well as he was able and instantly fell asleep himself. We jolted mournfully over the uneven pavements of Toulouse, I in the midst of these men who in the rain and the breaking day were about to take up again their dreary diurnal tasks, their red tape, their monotonous lives.

Morning after morning, greeted by the growl of the customs guard shaken out of sleep by his arrival, by the gruff irritability of clerk or inspector, one mail pilot or another got into this bus and was for the moment indistinguishable from these bureaucrats. But as the street lamps moved by, as the field drew nearer and nearer, the old omnibus rattling along lost little by little its reality and became a grey chrysalis from which one emerged transfigured.

Morning after morning a flyer sat here and felt of a sudden, somewhere inside the vulnerable man subjected to his neighbor's surliness, the stirring of the pilot of the Spanish and African mails, the birth of him who, three hours later, was to confront in the lightnings the dragon of the mountains; and who, four hours afterwards, having vanquished it, would be free to decide between a détour over the sea and a direct assault upon the Alcoy range, would be free to deal with storm, with mountain, with ocean.

And thus every morning each pilot before me, in his time, had been lost in the anonymity of daybreak beneath the dismal winter sky of Toulouse, and each one, transfigured by this old omnibus, had felt the birth within him of the sovereign who, five hours later, leaving behind him the rains and snows of the North, repudiating winter, had throttled down his motor and begun to drift earthward in the summer air beneath the shining sun of Alicante.

The old omnibus has vanished, but its austerity, its discomfort, still live in my memory. It was a proper symbol of the apprenticeship we had to serve before we might possess the stern joys of our craft. Everything about it was intensely serious. I remember three years later, though hardly ten words were spoken, learning in that bus of the death of Lécrivain, one of those hundred pilots who on a day or a night of fog have retired for eternity.

It was four in the morning, and the same silence was abroad when we heard the field manager, invisible in the darkness, address the inspector:

"Lécrivain didn't land at Casablanca last night."

"Ah!" said the inspector. "Ah?"

Torn from his dream he made an effort to wake up, to display his zeal, and added:

"Is that so? Couldn't he get through? Did he come back?"

And in the dead darkness of the omnibus the answer came: "No."

We waited to hear the rest, but no word sounded. And as the seconds fell its became more and more evident that that "no" would be followed by no further word, was eternal and without appeal, that Lécrivain not only had not landed at Casablanca but would never again land anywhere.

And so, at daybreak on the morning of my first flight with the mails, I went through the sacred rites of the craft, and I felt the self-confidence oozing out of me as I stared through the windows at the macadam shining and reflecting back the street lights. Over the pools of water I could see great palms of wind running. And I thought: "My first flight with the mails! Really, this is not my lucky day."

I raised my eyes and looked at the inspector. "Would you call this bad weather?" I asked.

He threw a weary glance out of the window. "Doesn't prove anything," he growled finally.

And I wondered how one could tell bad weather. The night before, with a single smile Guillaumet had wiped out all the evil omens with which the veterans overwhelmed us, but they came back into my memory. "I feel sorry for the man who doesn't know the whole line pebble by pebble, if he runs into a snow-storm. Oh, yes, I pity the fellow." Our elders, who had their prestige to think of, had all bobbed their heads solemnly and looked at us with embarrassing sympathy, as if they were pitying a flock of condemned sheep.

For how many of us had this old omnibus served as refuge in its day? Sixty? Eighty? I looked about me. Luminous points glowed in the darkness. Cigarettes punctuated the humble meditations of worn old clerks. How many of us had they escorted through the rain on a journey from which there was no coming back?

I heard them talking to one another in murmurs and whispers. They talked about illness, money, shabby domestic cares. Their talk painted the walls of the dismal prison in which these men had locked themselves up. And suddenly I had a vision of the face of destiny.

Old bureaucrat, my comrade, it is not you who are to blame. No one ever helped you to escape. You, like a termite, built your peace by blocking up with cement every chink and cranny through which the light might pierce. You rolled yourself up into a ball in your genteel security, in routine, in the stifling conventions of provincial life, raising a modest rampart against the winds and the tides and the stars. You have chosen not to be perturbed by great problems, having trouble enough to forget your own fate as man. You are not the dweller upon an errant planet and do not ask yourself questions to which there are no answers. You are a petty bourgeois of Toulouse. Nobody grasped you by the shoulder while there was still time. Now the clay of which you were shaped has dried and hardened, and naught in you will ever awaken the sleeping musician, the poet, the astronomer that possibly inhabited you in the beginning.

The squall has ceased to be a cause of my complaint. The magic of the craft has opened for me a world in which I shall confront, within two hours, the black dragons and the crowned crests of a coma of blue lightnings, and when night has fallen I, delivered, shall read my course in the stars.

Thus I went through my professional baptism and I began to fly the mails. For the most part the flights were without incident. Like sea-divers, we sank peacefully into the depths of our element.

Flying, in general, seemed to us easy. When the skies are filled with black vapors, when fog and sand and sea are confounded in a brew in which they become indistinguishable, when gleaming flashes wheel treacherously in these skyey swamps, the pilot purges himself of the phantoms at a single stroke. He lights his lamps. He brings sanity into his house as into a lonely cottage on a fearsome heath. And the crew travel a sort of submarine route in a lighted chamber.

Pilot, mechanic, and radio operator are shut up in what might be a laboratory. They are obedient to the play of dial-hands, not to the unrolling of the landscape. Out of doors the mountains are immersed in tenebrous darkness; but they are no longer mountains, they are invisible powers whose approach must be computed.

The operator sits in the light of his lamp, dutifully setting down figures; the mechanic ticks off points on his chart; the pilot swerves in response to the drift of the mountains as quickly as he sees that the summits he intends to pass on the left have deployed straight ahead of him in a silence and secrecy as of military preparations. And below on the ground the watchful radio men in their shacks take down submissively in their notebooks the dictation of their comrade in the air: "12:40 a.m. En route 230. All well."

So the crew fly on with no thought that they are in motion. Like night over the sea, they are very far from the earth, from towns, from trees. The motors fill the lighted chamber with a quiver that changes its substance. The clock ticks on. The dials, the radio lamps, the various hands and needles go through their invisible alchemy. From second to second these mysterious stirrings, a few muffled words, a concentrated tenseness, contribute to the end result. And when the hour is at hand the pilot may glue his forehead to the window with perfect assurance. Out of oblivion the gold has been smelted: there it gleams in the lights of the airport.

And yet we have all known flights when of a sudden, each for himself, it has seemed to us that we have crossed the border of the world of reality; when, only a couple of hours from port, we have felt ourselves more distant from it than we should feel if we were in India; when there has come a premonition of an incursion into a forbidden world whence it was going to be infinitely difficult to return.

Thus, when Mermoz first crossed the South Atlantic in a hydroplane, as day was dying he ran foul of the Black Hole region, off Africa. Straight ahead of him were the tails of tornadoes rising minute by minute gradually higher, rising as a wall is built; and then the night came down upon these preliminaries and swallowed them up; and when, an hour later, he slipped under the clouds, he came out into a fantastic kingdom.

Great black waterspouts had reared themselves seemingly in the immobility of temple pillars. Swollen at their tops, they were supporting the squat and lowering arch of the tempest, but through the rifts in the arch there fell slabs of light and the full moon sent her radiant beams between the pillars down upon the frozen tiles of the sea. Through these uninhabited ruins Mermoz made his way, gliding slantwise from one channel of light to the next, circling round those giant pillars in which there must have rumbled the upsurge of the sea, flying for four hours through these corridors of moonlight toward the exit from the temple. And this spectacle was so overwhelming that only after he had got through the Black Hole did Mermoz awaken to the fact that he had not been afraid.

I remember, for my part, another of those hours in which a pilot finds suddenly that he has slipped beyond the confines of this world. All that night the radio messages sent from the ports in the Sahara concerning our position had been inaccurate, and my radio operator, Néri, and I had been drawn out of our course. Suddenly, seeing the gleam of water at the bottom of a crevasse of fog, I tacked sharply in the direction of the coast; but it was by then impossible for us to say how long we had been flying towards the high seas. Nor were we certain of making the coast, for our fuel was probably low. And even so, once we had reached it we would still have to make port—after the moon had set.

We had no means of angular orientation, were already deafened, and were bit by bit growing blind. The moon like a pallid ember began to go out in the banks of fog. Overhead the sky was filling with clouds, and we flew thenceforth between cloud and fog in a world voided of all substance and all light. The ports that signaled us had given up trying to tell us where we were. "No bearings, no bearings," was all their message, for our voice reached them from everywhere and nowhere. With sinking hearts Néri and I leaned out, he on his side and I on mine, to see if anything, anything at all, was distinguishable in this void. Already our tired eyes were seeing things—errant signs, delusive flashes, phantoms.

And suddenly, when already we were in despair, low on the horizon a brilliant point was unveiled on our port bow. A wave of joy went through me. Néri leaned forward, and I could hear him singing. It could not but be the beacon of an airport, for after dark the whole Sahara goes black and forms a great dead expanse. That light twinkled for a space—and then went out! We had been steering for a star which was visible for a few minutes only, just before setting on the horizon between the layer of fog and the clouds.

Then other stars took up the game, and with a sort of dogged hope we set our course for each of them in turn. Each time that a light lingered a while, we performed the same crucial experiment. Néri would send his message to the airport at Cisneros: "Beacon in view. Put out your light and flash three times." And Cisneros would put out its beacon and flash three times while the hard light at which we gazed would not, incorruptible star, so much as wink. And despite our dwindling fuel we continued to nibble at the golden bait which each time seemed more surely the true light of a beacon, was each time a promise of a landing and of life—and we had each time to change our star.

And with that we knew ourselves to be lost in interplanetary space among a thousand inaccessible planets, we who sought only the one veritable planet, our own, that planet on which alone we should find our familiar countryside, the houses of our friends, our treasures.

On which alone we should find . . . Let me draw the picture that took shape before my eyes. It will seem to you childish; but even in the midst of danger a man retains his human concerns. I was thirsty and I was hungry. If we did find Cisneros we should re-fuel and carry on to Casablanca, and there we should come down in the cool of daybreak, free to idle the hours away. Néri and I would go into town. We would go to a little pub already open despite the early hour. Safe and sound, Néri and I would sit down at table and laugh at the night of danger as we ate our warm rolls and drank our bowls of coffee and hot milk. We would receive this matutinal gift at the hands of life. Even as an old peasant woman recognizes her God in a painted image, in a childish medal, in a chaplet, so life would speak to us in its humblest language in order that we understand. The joy of living, I say, was summed up for me in the remembered sensation of that first burning and aromatic swallow, that mixture of milk and coffee and bread by which men hold communion with tranquil pastures, exotic plantations, and golden harvests, communion with the earth. Amidst all these stars there was but one that could make itself significant for us by composing this aromatic bowl that was its daily gift at dawn. And from that earth of men, that earth docile to the reaping of grain and the harvesting of the grape, bearing its rivers asleep in their fields, its villages clinging to their hillsides, our ship was separated by astronomical distances. All the treasures of the world were summed up in a grain of dust now blown far out of our path by the very destiny itself of dust and of the orbs of night.

And Néri still prayed to the stars.

Suddenly he was pounding my shoulder. On the bit of paper he held forth impatiently to me I read: "All well. Magnificent news." I waited with beating heart while he scribbed the half-dozen words that were to save us. At last he put this grace of heaven into my hands.

It was dated from Casablanca, which we had left the night before. Delayed in transmission, it had suddenly found us more than a thousand miles away, suspended between cloud and fog, lost at sea. It was sent by the government representative at the airport. And it said: "Monsieur de Saint-Exupéry, I am obliged to recommend that you be disciplined at Paris for having flown too close to the hangars on leaving Casablanca."

It was true that I had done this. It was also true that this man was performing his duty with irritability. I should have been humiliated if this reproach had been addressed to me in an airport. But it reached me where it had no right to reach me. Among these too rare stars, on this bed of fog, in this menacing savor of the sea, it burst like a detonation. Here we were with our fate in our hands, the fate of the mails and of the ship; we had trouble enough to try to keep alive; and this man was purging his petty rancor against us.

But Néri and I were far from nettled. What we felt was a vast and sudden jubilation. Here it was we who were masters, and this man was letting us know it. The impudent little corporal! not to have looked at our stripes and seen that we had been promoted captain! To intrude into our musings when we were solemnly taking our constitutional between Sagittarius and the Great Bear! When the only thing we could be concerned with, the only thing of our order of magnitude, was this appointment we were missing with the moon!

The immediate duty, the only duty of the planet whence this man's message came, was to furnish us accurate figures for our computations among the stars. And its figures had been false. This being so, the planet had only to hold its tongue. Néri scribbed: "Instead of wasting their time with this nonsense they would do better to haul us back to Cisneros, if they can." By "they" he meant all the peoples of the globe, with their parliaments, their senates, their navies, their armies, their emperors. We re-read the message from that man mad enough to imagine that he had business with us, and tacked in the direction of Mercury.

It was by the purest chance that we were saved. I had given up all thought of making Cisneros and had set my course at right angles to the coast-line in the hope that thus we might avoid coming down at sea when our fuel ran out. Meanwhile however I was in the belly of a dense fog so that even with land below it was not going to be easy to set the ship down. The situation was so clear that already I was shrugging my shoulders ruefully when Néri passed me a second message which, an hour earlier, would have been our salvation. "Cisneros," it said, "has deigned to communicate with us. Cisneros says, '216 doubtful.' " Well, that helped.

Cisneros was no longer swallowed up in space, it was actually out there on our left, almost within reach. But how far away? Néri and I talked it over briefly, decided it was too late to try for it (since that might mean missing the coast), and Néri replied: "Only one hour fuel left continuing on 93."

But the airports one by one had been waking each other up. Into our dialogue broke the voices of Agadir, Casablanca, Dakar. The radio stations at each of these towns had warned the airports and the ports had flashed the news to our comrades. Bit by bit they were gathering round us as round a sick-bed. Vain warmth, but human warmth after all. Helpless concern, but affectionate at any rate.

And suddenly into this conclave burst Toulouse, the headquarters of the Line three thousand miles away, worried along with the rest. Toulouse broke in without a word of greeting, simply to say sharply: "Your reserve tanks bigger than standard. You have two hours fuel left. Proceed to Cisneros."

There is no need of nights like the one just described to make the airline pilot find new meanings in old appearances. The scene that strikes the passenger as commonplace is from the very moment of taking off animated with a powerful magic for the crew. It is the duty of the ship's captain to make port, cost what it may. The sight of massing clouds is no mere spectacle to him: it is a matter of concern to his physical being, and to his mind it means a set of problems. Before he is off the ground he has taken its measure, and between him and it a bond is formed which is a veritable language.

There is a peak ahead, still distant. The pilot will not reach it before another hour of flight in the night. What is to be the significance of that peak? On a night of full moon it will be a useful landmark. In fainter moonglow it will be a bit of wreckage strewn in shadow, dangerous, but marked clearly enough by the lights of villages. But if the pilot flies blind, has bad luck in correcting his drift, is dubious about his position, that peak begins to stir with a strange life and its threat fills the breadth of the night sky in the same way as a single mine, drifting at the will of the current, can render the whole of the ocean a danger.

The face of the sea is as variable as that of the earth. To passengers, the storm is invisible. Seen from a great height, the waves have no relief and the packets of fog have no movements. The surface of the sea appears to be covered with great white motionless palm-trees, palms marked with ribs and seams stiff in a sort of frost. The sea is like a splintered mirror. But the hydroplane pilot knows there is no landing here.

The hours during which a man flies over this mirror are hours in which there is no assurance of the possession of anything in the world. These palms beneath the plane are so many poisoned flowers. And even when the flight is an easy one, made under a shining sun, the pilot navigating at some point on the line is not gazing upon a scene. These colors of earth and sky, these traces of wind over the face of the sea, these clouds golden in the afterglow, are not objects of the pilot's admiration, but of his cogitation. He looks to them to tell him the direction of the wind or the progress of the storm, and the quality of the night to come.

Even as the peasant strolling about his domain is able to foresee in a thousand signs the coming of the spring, the threat of frost, a promise of rain, so all that happens in the sky signals to the pilot the oncoming snow, the expectancy of fog, or the peace of a blessed night. The machine which at first blush seems a means of isolating man from the great problems of nature, actually plunges him more deeply into them. As for the peasant so for the pilot, dawn and twilight become events of consequence. His essential problems are set him by the mountain, the sea, the wind. Alone before the vast tribunal of the tempestuous sky, the pilot defends his mails and debates on terms of equality with those three elemental divinities.

The mail pouches for which he is responsible are stowed away in the after hold. They constitute the dogma of the religion of his craft, the torch which, in this aerial race, is passed from runner to runner. What matter though they hold but the scribblings of tradesmen and nondescript lovers. The interests which dictated them may very well not be worth the embrace of man and storm; but I know what they become once they have been entrusted to the crew, taken over, as the phrase is. The crew care not a rap for banker or tradesman. If, some day, the crew are hooked by a cliff it will not have been in the interest of tradespeople that they will have died, but in obedience to orders which ennoble the sacks of mail once they are on board ship.

What concerns us is not even the orders—it is the men they cast in their mould.

# II

# *The Men*

Mermoz is one airline pilot, and Guillaumet another, of whom I shall write briefly in order that you may see clearly what I mean when I say that in the mould of this new profession a new breed of men has been cast.

## I

A handful of pilots, of whom Mermoz was one, surveyed the Casablanca-Dakar line across the territory inhabited by the refractory tribes of the Sahara. Motors in those days being what they were, Mermoz was taken prisoner one day by the Moors. The tribesmen were unable to make up their minds to kill him, kept him a captive a fortnight, and he was eventually ransomed. Whereupon he continued to fly over the same territory.

When the South American line was opened up Mermoz, ever the pioneer, was given the job of surveying the division between Buenos Aires and Santiago de Chile. He who had flung a bridge over the Sahara was now to do the same over the Andes. They had given him a plane whose absolute ceiling was sixteen thousand feet and had asked him to fly it over a mountain range that rose more than twenty thousand feet into the air. His job was to search for gaps in the Cordilleras. He who had studied the face of the sands was now to learn the contours of the peaks, those crags whose scarfs of snow flutter restlessly in the winds, whose surfaces are bleached white in the storms, whose blustering gusts sweep through the narrow walls of their rocky corridors and force the pilot to a sort of hand-to-hand combat. Mermoz enrolled in this war in complete ignorance of his adversary, with no notion at all of the chances of coming forth alive from battle with this enemy. His job was to "try out" for the rest of us. And, "trying out" one day, he found himself prisoner of the Andes.

Mermoz and his mechanic had been forced down at an altitude of twelve thousand feet on a table-land at whose edges the mountain dropped sheer on all sides. For two mortal days they hunted a way off this plateau. But they were trapped. Everywhere the same sheer drop. And so they played their last card.

Themselves still in it, they sent the plane rolling and bouncing down an incline over the rocky ground until it reached the precipice, went off into air, and dropped. In falling, the plane picked up enough speed to respond to the controls. Mermoz was able to tilt its nose in the direction of a peak, sweep over the peak, and, while the water spurted through all the pipes burst by the night frost, the ship already disabled after only seven minutes of flight, he saw beneath him like a promised land the Chilean plain.

And the next day he was at it again.

When the Andes had been thoroughly explored and the technique of the crossings perfected, Mermoz turned over this section of the line to his friend Guillaumet and set out to explore the night. The lighting of our airports had not yet been worked out. Hovering in the pitch black night, Mermoz would land by the faint glimmer of three gasoline flares lined up at one end of the field. This trick, too, he taught us, and then, having tamed the night, he tried the ocean. He was the first, in 1931, to carry the mails in four days from Toulouse to Buenos Aires. On his way home he had engine trouble over a stormy sea in mid-Atlantic. A passing steamer picked him up with his mails and his crew.

Pioneering thus, Mermoz had cleared the desert, the mountains, the night, and the sea. He had been forced down more than once in desert, in mountain, in night, and in sea. And each time that he got safely home, it was but to start out again. Finally, after a dozen years of service, having taken off from Dakar bound for Natal, he radioed briefly that he was cutting off his rear right-hand engine. Then silence.

There was nothing particularly disturbing in this news. Nevertheless, when ten minutes had gone by without report there began for every radio station on the South Atlantic line, from Paris to Buenos Aires, a period of anxious vigil. It would be ridiculous to worry over someone ten minutes late in our day-to-day existence, but in the air-mail service ten minutes can be pregnant with meaning. At the heart of this dead slice of time an unknown event is locked up. Insignificant, it may be; a mishap, possibly: whatever it is, the event has taken place. Fate has pronounced a decision from which there is no appeal. An iron hand has guided a crew to a sea-landing that may have been safe and may have been disastrous. And long hours must go by before the decision of the gods is made known to those who wait.

We waited. We hoped. Like all men at some time in their lives we lived through that inordinate expectancy which like a fatal malady grows from minute to minute harder to bear. Even before the hour sounded, in our hearts many among us were already sitting up with the dead. All of us had the same vision before our eyes. It was a vision of a cockpit still inhabited by living men; but the pilot's hands were telling him very little now, and the world in which he groped and fumbled was a world he did not recognize. Behind him, in the glimmer of the cabin light, a shapeless uneasiness floated. The crew moved to and fro, discussed their plight, feigned sleep. A restless slumber it was, like the stirring of drowned men. The only element of sanity, of intelligibility, was the whirring of the three engines with its reassuring evidence that time still existed for them.

We were haunted for hours by this vision of a plane in distress. But the hands of the clock were going round and little by little it began to grow late. Slowly the truth was borne in upon us that our comrades would never return, that they were sleeping in that South Atlantic whose skies they had so often ploughed. Mermoz had done this job and slipped away to rest, like a gleaner who, having carefully bound his sheaf, lies down in the field to sleep.

When a pilot dies in the harness his death seems something that inheres in the craft itself, and in the beginning the hurt it brings is perhaps less than the pain sprung of a different death. Assuredly he has vanished, has undergone his ultimate mutation; but his presence is still not missed as deeply as we might miss bread. For in this craft we take it for granted that we shall meet together only rarely.

Airline pilots are widely dispersed over the face of the world. They land alone at scattered and remote airports, isolated from each other rather in the manner of sentinels between whom no words can be spoken. It needs the accident of journeyings to bring together here or there the dispersed members of this great professional family.

Round the table in the evening, at Casablanca, at Dakar, at Buenos Aires, we take up conversations interrupted by years of silence, we resume friendships to the accompaniment of buried memories. And then we are off again.

Thus is the earth at once a desert and a paradise, rich in secret hidden gardens, gardens inaccessible, but to which the craft leads us ever back, one day or another. Life may scatter us and keep us apart; it may prevent us from thinking very often of one another; but we know that our comrades are somewhere "out there"—where, one can hardly say—silent, forgotten, but deeply faithful. And when our path crosses theirs, they greet us with such manifest joy, shake us so gaily by the shoulders! Indeed we are accustomed to waiting.

Bit by bit, nevertheless, it comes over us that we shall never again hear the laughter of our friend, that this one garden is forever locked against us. And at that moment begins our true mourning, which, though it may not be rending, is yet a little bitter. For nothing, in truth, can replace that companion. Old friends cannot be created out of hand. Nothing can match the treasure of common memories, of trials endured together, of quarrels and reconciliations and generous emotions. It is idle, having planted an acorn in the morning, to expect that afternoon to sit in the shade of the oak.

So life goes on. For years we plant the seed, we feel ourselves rich; and then come other years when time does its work and our plantation is made sparse and thin. One by one, our comrades slip away, deprive us of their shade.

This, then, is the moral taught us by Mermoz and his kind. We understand better, because of him, that what constitutes the dignity of a craft is that it creates a fellowship, that it binds men together and fashions for them a common language. For there is but one veritable problem—the problem of human relations.

We forget that there is no hope of joy except in human relations. If I summon up those memories that have left with me an enduring savor, if I draw up the balance sheet of the hours in my life that have truly counted, surely I find only those that no wealth could have procured me. True riches cannot be bought. One cannot buy the friendship of a Mermoz, of a companion to whom one is bound forever by ordeals suffered in common. There is no buying the night flight with its hundred thousand stars, its serenity, its few hours of sovereignty. It is not money that can procure for us that new vision of the world won through hardship—those trees, flowers, women, those treasures made fresh by the dew and color of life which the dawn restores to us, this concert of little things that sustain us and constitute our compensation.

Nor that night we lived through in the land of the unconquered tribes of the Sahara, which now floats into my memory.

Three crews of Aéropostale men had come down at the fall of day on the Rio de Oro coast in a part of the Sahara whose denizens acknowledge no European rule. Riguelle had landed first, with a broken connecting rod. Bourgat had come along to pick up Riguelle's crew, but a minor accident had nailed him to earth. Finally, as night was beginning to fall, I arrived. We decided to salvage Bourgat's ship, but we should have to spend the night and do the job of repair by daylight.

Exactly on this spot two of our comrades, Gourp and Erable, had been murdered by the tribesmen a year earlier. We knew that a raiding party of three hundred rifles was at this very moment encamped somewhere near by, round Cape Bojador. Our three landings had been visible from a great distance and the Moors must have seen us. We began a vigil which might turn out to be our last.

Altogether, there were about ten of us, pilots and mechanics, when we made ready for the night. We unloaded five or six wooden cases of merchandise out of the hold, emptied them, and set them about in a circle. At the deep end of each case, as in a sentry-box, we set a lighted candle, its flame poorly sheltered from the wind. So in the heart of the desert, on the naked rind of the planet, in an isolation like that of the beginnings of the world, we built a village of men.

Sitting in the flickering light of the candles on this kerchief of sand, on this village square, we waited in the night. We were waiting for the rescuing dawn—or for the Moors. Something, I know not what, lent this night a savor of Christmas. We told stories, we joked, we sang songs. In the air there was that slight fever that reigns over a gaily prepared feast. And yet we were infinitely poor. Wind, sand, and stars. The austerity of Trappists. But on this badly lighted cloth, a handful of men who possessed nothing in the world but their memories were sharing invisible riches.

We had met at last. Men travel side by side for years, each locked up in his own silence or exchanging those words which carry no freight—till danger comes. Then they stand shoulder to shoulder. They discover that they belong to the same family. They wax and bloom in the recognition of fellow beings. They look at one another and smile. They are like the prisoner set free who marvels at the immensity of the sea.

Happiness! It is useless to seek it elsewhere than in this warmth of human relations. Our sordid interests imprison us within their walls. Only a comrade can grasp us by the hand and haul us free.

And these human relations must be created. One must go through an apprenticeship to learn the job. Games and risk are a help here. When we exchange manly handshakes, compete in races, join together to save one of us who is in trouble, cry aloud for help in the hour of danger—only then do we learn that we are not alone on earth.

Each man must look to himself to teach him the meaning of life. It is not something discovered: it is something moulded. These prison walls that this age of trade has built up round us, we can break down. We can still run free, call to our comrades, and marvel to hear once more, in response to our call, the pathetic chant of the human voice.

## II

Guillaumet, old friend, of you too I shall say a few words. Be sure that I shall not make you squirm with any clumsy vaunting of your courage and your professional valor. In telling the story of the most marvelous of your adventures, I am after something quite different.

There exists a quality which is nameless. It may be gravity, but the word does not satisfy me, for the quality I have in mind can be accompanied by the most cheerful gaiety. It is the quality of the carpenter face to face with his block of wood. He handles it, he takes its measure. Far from treating it frivolously, he summons all his professional virtues to do it honor.

I once read, Guillaumet, a tale in which your adventure was celebrated. I have an old score to settle with the infidel who wrote it. You were described as abounding in the witty sallies of the street arab, as if courage consisted in demeaning oneself to schoolboy banter in the midst of danger and the hour of death. The man did not know you, Guillaumet. You never felt the need of cheapening your adversaries before confronting them. When you saw a foul storm you said to yourself, "Here is a foul storm." You accepted it, and you took its measure.

These pages, Guillaumet, written out of my memory, are addressed in homage to you.

It was winter and you had been gone a week over the Andes. I had come up from farthest Patagonia to join Deley at Mendoza. For five days the two of us, each in his plane, had ransacked the mountains unavailingly. Two ships! It seemed to us that a hundred squadrons navigating for a hundred years would not have been enough to explore that endless, cloud-piercing range. We had lost all hope. The very smugglers themselves, bandits who would commit a crime for a five-peso note, refused to form a rescue party out of fear of those counterforts. "We should surely die," they said; "the Andes never give up a man in winter."

And when Deley and I landed at Santiago, the Chilean officers also advised us to give you up. "It is mid-winter," they said; "even if your comrade survived the landing, he cannot have survived the night. Night in those passes changes a man into ice."

And when, a second time, I slipped between the towering walls and giant pillars of the Andes, it seemed to me I was no longer seeking, but was now sitting up with, your body in the silence of a cathedral of snow.

You had been gone a week, I say, and I was lunching between

flights in a restaurant in Mendoza when a man stuck his head in the door and called out:

"They've found Guillaumet!"

All the strangers in the restaurant embraced.

Ten minutes later I was off the ground, carrying two mechanics, Lefebvre and Abri. Forty minutes later I had landed alongside a road, having recognized from the air, I know not by what sign, the car in which you were being brought down from San Rafael. I remember that we cried like fools; we put our arms about a living Guillaumet, resuscitated, the author of his own miracle. And it was at that moment that you pronounced your first intelligible sentence, a speech admirable in its human pride:

"I swear that what I went through, no animal would have gone through."

Later, you told us the story. A storm that brought fifteen feet of snow in forty-eight hours down on the Chilean slope had bottled up all space and sent every other mail pilot back to his starting point. You, however, had taken off in the hope of finding a rift in the sky. You found this rift, this trap, a little to the south, and now, at twenty thousand feet, the ceiling of clouds being a couple of thousand feet below you and pierced by only the highest peaks, you set your course for Argentina.

Down currents sometimes fill pilots with a strange uneasiness. The engines run on, but the ship seems to be sinking. You jockey to hold your altitude: the ship loses speed and goes mushy. And still you sink. So you give it up, afraid that you may have jockeyed too much; and you let yourself drift to right or left, striving to put at your back a favorable peak, that is, a peak off which the winds rebound as off a springboard.

And yet you go on sinking. The whole sky seems to be coming down on you. You begin to feel like the victim of some cosmic accident. You cannot land anywhere, and you try in vain to turn round and fly back into those zones where the air, as dense and solid as a pillar, had held you up. That pillar has melted away. Everything here is rotten and you slither about in a sort of universal decomposition while the cloud-bank rises apathetically, reaches your level, and swallows you up.

"It almost had me in a corner once," you explained, "but I still wasn't sure I was caught. When you get up above the clouds you run into those down currents that seem to be perfectly stationary for the simple reason that in that very high altitude they never stop flowing. Everything is queer in the upper range."

And what clouds!

"As soon as I felt I was caught I dropped the controls and grabbed my seat for fear of being flung out of the ship. The jolts were so terrible that my leather harness cut my shoulders and was ready to snap. And what with the frosting on the panes, my artificial horizon was invisible and the wind rolled me over and over like a hat in a road from eighteen thousand feet down to ten.

"At ten thousand I caught a glimpse of a dark horizontal blot that helped me right the ship. It was a lake, and I recognized it as what they call Laguna Diamante. I remembered that it lay at the bottom of a funnel, and that one flank of the funnel, a volcano called Maipu, ran up to about twenty thousand feet.

"There I was, safe out of the clouds; but I was still blinded by the thick whirling snow and I had to hang on to my lake if I wasn't to crash into one of the sides of the funnel. So down I went, and I flew round and round the lake, about a hundred and fifty feet above it, until I ran out of fuel. After two hours of this, I set the ship down on the snow—and over on her nose she went.

"When I dragged myself clear of her I stood up. The wind knocked me down. I stood up again. Over I went a second time. So

I crawled under the cockpit and dug me out a shelter in the snow. I pulled a lot of mail sacks round me, and there I lay for two days and two nights. Then the storm blew over and I started to walk my way out. I walked for five days and four nights."

But what was there left of you, Guillaumet? We had found you again, true; but burnt to a crisp, but shriveled, but shrunken into an old woman. That same afternoon I flew you back to Mendoza, and there the cool white sheets flowed like a balm down the length of your body.

They were not enough, though. Your own foundered body was an encumbrance: you turned and twisted in your sleep, unable to find lodgment for it. I stared at your face: it was splotched and swollen, like an overripe fruit that has been repeatedly dropped on the ground.

You were dreadful to see, and you were in misery, for you had lost the beautiful tools of your works: your hands were numb and useless, and when you sat up on the edge of your bed to draw a free breath, your frozen feet hung down like two dead weights. You had not even finished your long walk back, you were still panting; and when you turned and stirred on the pillow in search of peace, a procession of images that you could not escape, a procession waiting impatiently in the wings, moved instantly into action under your skull. Across the stage of your skull it moved, and for the twentieth time you fought once more the battle against these enemies that rose up out of their ashes.

I filled you with herb-teas.

"Drink, old fellow."

"You know . . . what amazed me . . ."

Boxer victorious, but punch-drunk and scarred with blows, you were re-living your strange adventure. You could divest yourself of it only in scraps. And as you told your dark tale, I could see you trudging without ice-axe, without ropes, without provisions, scaling cols fifteen thousand feet in the air, crawling on the faces of vertical walls, your hands and feet and knees bleeding in a temperature twenty degrees below zero.

Voided bit by bit of your blood, your strength, your reason, you went forward with the obstinacy of an ant, retracing your steps to go round an obstacle, picking yourself up after each fall to earth, climbing slopes that led to abysses, ceaselessly in motion and never asleep, for had you slept, from that bed of snow you would never have risen. When your foot slipped and you went down, you were up again in an instant, else had you been turned into stone. The cold was petrifying you by the minute, and the price you paid for taking a moment too much of rest, when you fell, was the agony of revivifying dead muscles in your struggle to rise to your feet.

You resisted temptation. "Amid snow," you told me, "a man loses his instinct of self-preservation. After two or three or four days of tramping, all you think about is sleep. I would long for it; but then I would say to myself, 'If my wife still believes I am alive, she must believe that I am on my feet. The boys all think I am on my feet. They have faith in me. And I am a skunk if I don't go on.' "

So you tramped on; and each day you cut out a bit more of the opening of your shoes so that your swelling and freezing feet might have room in them.

You confided to me this strange thing:

"As early as the second day, you know, the hardest job I had was to force myself not to think. The pain was too much, and I was really up against it too hard. I had to forget that, or I shouldn't have had the heart to go on walking. But I didn't seem able to control my mind. It kept working like a turbine. Still, I could more or less choose what I was to think about. I tried to stick to some film I'd seen, or book I'd read. But the film and the book would go through my mind like lightning, and I'd be back where I was, in the snow. It never failed. So I would think about other things. . . ."

There was one time, however, when, having slipped, and finding yourself stretched flat on your face in the snow, you threw in your hand. You were like a boxer emptied of all passion by a single blow, lying and listening to the seconds drop one by one into a distant universe, until the tenth second fell and there was no appeal.

"I've done my best and I can't make it. Why go on?" All that you had to do in the world to find peace was to shut your eyes. So little was needed to blot out that world of crags and ice and snow. Let drop those miraculous eyelids and there was an end of blows, of stumbling falls, of torn muscles and burning ice, of that burden of life you were dragging along like a worn-out ox, a weight heavier than any wain or cart.

Already you were beginning to taste the relief of this snow that had now become an insidious poison, this morphia that was filling you with beatitude. Life crept out of your extremities and fled to collect round your heart while something gentle and precious snuggled in close at the centre of your being. Little by little your consciousness deserted the distant regions of your body, and your body, that beast now gorged with suffering, lay ready to participate in the indifference of marble.

Your very scruples subsided. Our cries ceased to reach you, or, more accurately, changed for you into dream-cries. You were happy now, able to respond by long confident dream-strides that carried you effortlessly towards the enchantment of the plains below. How smoothly you glided into this suddenly merciful world! Guillaumet, you miser! You had made up your mind to deny us your return, to take your pleasures selfishly without us among your white angels in the snows. And then remorse floated up from the depth of your consciousness. The dream was spoilt by the irruption of bothersome details. "I thought of my wife. She would be penniless if she couldn't collect the insurance. Yes, but the company . . ."

When a man vanishes, his legal death is postponed for four years. This awful detail was enough to blot out the other visions. You were lying face downward on a bed of snow that covered a steep mountain slope. With the coming of summer your body would be washed with this slush down into one of the thousand crevasses of the Andes. You knew that. But you also knew that some fifty yards away a rock was jutting up out of the snow. "I thought, if I get up I may be able to reach it. And if I can prop myself up against the rock, they'll find me there next summer."

Once you were on your feet again, you tramped two nights and three days. But you did not then imagine that you would go on much longer:

"I could tell by different signs that the end was coming. For instance, I had to stop every two or three hours to cut my shoes open a bit more and massage my swollen feet. Or maybe my heart would be going too fast. But I was beginning to lose my memory. I had been going on a long time when suddenly I realized that every time I stopped I forgot something. The first time it was a glove. And it was cold! I had put it down in front of me and had forgotten to pick it up. The next time it was my watch. Then my knife. Then my compass. Each time I stopped I stripped myself of something vitally important. I was becoming my own enemy! And I can't tell you how it hurt me when I found that out.

"What saves a man is to take a step. Then another step. It is always the same step, but you have to take it."

"I swear that what I went through, no animal would have gone through." This sentence, the noblest ever spoken, this sentence that defines man's place in the universe, that honors him, that re-establishes the true hierarchy, floated back into my thoughts. Finally you fell asleep. Your consciousness was abolished; but forth from this dismantled, burnt, and shattered body it was to be born again like a flower put forth gradually by the species which itself is born of the luminous pulp of the stars. The body, we may say, then, is but an honest tool, the body is but a servant. And it was in these words, Guillaumet, that you expressed your pride in the honest tool:

"With nothing to eat, after three days on my feet . . . well . . . my heart wasn't going any too well. I was crawling along the side of a sheer wall, hanging over space, digging and kicking out pockets in the ice so that I could hold on, when all of a sudden my heart conked. It hesitated. Started up again. Beat crazily. I said to myself, 'If it hesitates a moment too long, I drop.' I stayed still and listened to myself. Never, never in my life have I listened as carefully to a motor as I listened to my heart, me hanging there. I said to it: 'Come on, old boy. Go to work. Try beating a little.' That's good stuff my heart is made of. It hesitated, but it went on. You don't know how proud I was of that heart."

As I said, in that room in Mendoza where I sat with you, you fell finally into an exhausted sleep. And I thought: If we were to talk to him about his courage, Guillaumet would shrug his shoulders. But it would be just as false to extol his modesty. His place is far beyond that mediocre virtue.

If he shrugs his shoulders, it is because he is no fool. He knows that once men are caught up in an event they cease to be afraid.

Only the unknown frightens men. But once a man has faced the unknown, that terror becomes the known.

Especially if it is scrutinized with Guillaumet's lucid gravity. Guillaumet's courage is in the main the product of his honesty. But even this is not his fundamental quality. His moral greatness consists in his sense of responsibility. He knew that he was responsible for himself, for the mails, for the fulfilment of the hopes of his comrades. He was holding in his hands their sorrow and their joy. He was responsible for that new element which the living were constructing and in which he was a participant. Responsible, in as much as his work contributed to it, for the fate of those men.

Guillaumet was one among those bold and generous men who had taken upon themselves the task of spreading their foliage over bold and generous horizons. To be a man is, precisely, to be responsible. It is to feel shame at the sight of what seems to be unmerited misery. It is to take pride in a victory won by one's comrades. It is to feel, when setting one's stone, that one is contributing to the building of the world.

There is a tendency to class such men with toreadors and gamblers. People extol their contempt for death. But I would not give a fig for anybody's contempt for death. If its roots are not sunk deep in an acceptance of responsibility, this contempt for death is the sign either of an impoverished soul or of youthful extravagance.

I once knew a young suicide. I cannot remember what disappointment in love it was which induced him to send a bullet carefully into his heart. I have no notion what literary temptation he had succumbed to when he drew on a pair of white gloves before the shot. But I remember having felt, on learning of this sorry show, an impression not of nobility but of lack of dignity. So! Behind that attractive face, beneath that skull which should have been a treasure chest, there had been nothing, nothing at all. Unless it was the vision of some silly little girl indistinguishable from the rest.

And when I heard of this meagre destiny, I remembered the death of a man. He was a gardener, and he was speaking on his deathbed: "You know, I used to sweat sometimes when I was digging. My rheumatism would pull at my leg, and I would damn myself for a slave. And now, do you know, I'd like to spade and spade. It's beautiful work. A man is free when he is using a spade. And besides, who is going to prune my trees when I am gone?"

That man was leaving behind him a fallow field, a fallow planet. He was bound by ties of love to all cultivable land and to all the trees of the earth. There was a generous man, a prodigal man, a nobleman! There was a man who, battling against death in the name of his Creation, could like Guillaumet be called a man of courage!

# III

## *The Tool*

And now, having spoken of the men born of the pilot's craft, I shall say something about the tool with which they work—the airplane. Have you looked at a modern airplane? Have you followed from year to year the evolution of its lines? Have you ever thought, not only about the airplane but about whatever man builds, that all of man's industrial efforts, all his computations and calculations, all the nights spent over working draughts and blueprints, invariably culminate in the production of a thing whose sole and guiding principle is the ultimate principle of simplicity?

It is as if there were a natural law which ordained that to achieve this end, to refine the curve of a piece of furniture, or a ship's keel, or the fuselage of an airplane, until gradually it partakes of the elementary purity of the curve of a human breast or shoulder, there must be the experimentation of several generations of craftsmen. In anything at all, perfection is finally attained not when there is no longer anything to add, but when there is no longer anything to take away, when a body has been stripped down to its nakedness.

It results from this that perfection of invention touches hands with absence of invention, as if that line which the human eye will follow with effortless delight were a line that had not been invented but simply discovered, had in the beginning been hidden by nature and in the end been found by the engineer. There is an ancient myth about the image asleep in the block of marble until it is carefully disengaged by the sculptor. The sculptor must himself feel that he is not so much inventing or shaping the curve of breast or shoulder as delivering the image from its prison.

In this spirit do engineers, physicists concerned with thermodynamics, and the swarm of preoccupied draughtsmen tackle their work. In appearance, but only in appearance, they seem to be polishing surfaces and refining away angles, easing this joint or stabilizing that wing, rendering these parts invisible, so that in the end there is no longer a wing hooked to a framework but a form flawless in its perfection, completely disengaged from its matrix, a sort of spontaneous whole, its parts mysteriously fused together and resembling in their unity a poem.

Meanwhile, startling as it is that all visible evidence of invention should have been refined out of this instrument and that there should be delivered to us an object as natural as a pebble polished by the waves, it is equally wonderful that he who uses this instrument should be able to forget that it is a machine.

There was a time when a flyer sat at the centre of a complicated works. Flight set us factory problems. The indicators that oscillated on the instrument panel warned us of a thousand dangers. But in the machine of today we forget that motors are whirring: the motor, finally, has come to fulfil its function, which is to whirr as a heart beats—and we give no thought to the beating of our heart. Thus, precisely because it is perfect the machine dissembles its own existence instead of forcing itself upon our notice.

And thus, also, the realities of nature resume their pride of place. It is not with metal that the pilot is in contact. Contrary to the vulgar illusion, it is thanks to the metal, and by virtue of it, that the pilot rediscovers nature. As I have already said, the machine does not isolate man from the great problems of nature but plunges him more deeply into them.

Numerous, nevertheless, are the moralists who have attacked the machine as the source of all the ills we bear, who, creating a fictitious dichotomy, have denounced the mechanical civilization as the enemy of the spiritual civilization.

If what they think were really so, then indeed we should have to despair of man, for it would be futile to struggle against this new advancing chaos. The machine is certainly as irresistible in its advance as those virgin forests that encroach upon equatorial domains. A congeries of motives prevents us from blowing up our spinning mills and reviving the distaff. Gandhi had a try at this sort of revolution: he was as simple-minded as a child trying to empty the sea on to the sand with the aid of a teacup.

It is hard for me to understand the language of these pseudo-dreamers. What is it makes them think that the ploughshare torn from the bowels of the earth by perforating machines, forged, tempered, and sharpened in the roar of modern industry, is nearer to man than any other tool of steel? By what sign do they recognize the inhumanity of the machine?

Have they ever really asked themselves this question? The central struggle of men has ever been to understand one another, to join together for the common weal. And it is this very thing that the machine helps them to do! It begins by annihilating time and space.

To me, in France, a friend speaks from America. The energy that brings me his voice is born of dammed-up waters a thousand miles from where he sits. The energy I burn up in listening to him is dispensed in the same instant by a lake formed in the River Yser which, four thousand miles from him and five hundred from me, melts like snow in the action of the turbines. Transport of the mails, transport of the human voice, transport of flickering pictures —in this century as in others our highest accomplishments still have the single aim of bringing men together. Do our dreamers hold that the invention of writing, of printing, of the sailing ship, degraded the human spirit?

It seems to me that those who complain of man's progress confuse ends with means. True, that man who struggles in the unique hope of material gain will harvest nothing worth while. But how can anyone conceive that the machine is an end? It is a tool. As much a tool as is the plough. The microscope is a tool. What disservice do we do the life of the spirit when we analyze the universe through a tool created by the science of optics, or seek to bring together those who love one another and are parted in space?

"Agreed!" my dreamers will say, "but explain to us why it is that a decline in human values has accompanied the rise of the machine?" Oh, I miss the village with its crafts and its folksongs as much as they do! The town fed by Hollywood seems to me.

too, impoverished despite its electric street lamps. I quite agree that men lose their creative instincts when they are fed thus without raising a hand. And I can see that it is tempting to accuse industry of this evil.

But we lack perspective for the judgment of transformations that go so deep. What are the hundred years of the history of the machine compared with the two hundred thousand years of the history of man? It was only yesterday that we began to pitch our camp in this country of laboratories and power stations, that we took possession of this new, this still unfinished, house we live in. Everything round us is new and different—our concerns, our working habits, our relations with one another.

Our very psychology has been shaken to its foundations, to its most secret recesses. Our notions of separation, absence, distance, return, are reflections of a new set of realities, though the words themselves remain unchanged. To grasp the meaning of the world of today we use a language created to express the world of yesterday. The life of the past seems to us nearer our true natures, but only for the reason that it is nearer our language.

Every step on the road of progress takes us farther from habits which, as the life of man goes, we had only recently begun to acquire. We are in truth emigrants who have not yet founded our homeland. We Europeans have become again young peoples, without tradition or language of our own. We shall have to age somewhat before we are able to write the folksongs of a new epoch.

Young barbarians still marveling at our new toys—that is what we are. Why else should we race our planes, give prizes to those who fly highest, or fastest? We take no heed to ask ourselves why we race: the race itself is more important than the object.

And this holds true of other things than flying. For the colonial soldier who founds an empire, the meaning of life is conquest. He despises the colonist. But was not the very aim of his conquest the settling of this same colonist?

In the enthusiasm of our rapid mechanical conquests we have overlooked some things. We have perhaps driven men into the service of the machine, instead of building machinery for the service of man. But could anything be more natural? So long as we were engaged in conquest, our spirit was the spirit of conquerors. The time has now come when we must be colonists, must make this house habitable which is still without character.

Little by little the machine will become part of humanity. Read the history of the railways in France, and doubtless elsewhere too: they had all the trouble in the world to tame the people of our villages. The locomotive was an iron monster. Time had to pass before men forgot what it was made of. Mysteriously, life began to run through it, and now it is wrinkled and old. What is it today for the villager except a humble friend who calls every evening at six?

The sailing vessel itself was once a machine born of the calculations of engineers, yet it does not disturb our philosophers. The sloop took its place in the speech of men. There is a poetry of sailing as old as the world. There have always been seamen in recorded time. The man who assumes that there is an essential difference between the sloop and the airplane lacks historic perspective.

Every machine will gradually take on this patina and lose its identity in its function.

Air and water, and not machinery, are the concern of the hydroplane pilot about to take off. The motors are running free and the plane is already ploughing the surface of the sea. Under the dizzying whirl of the scythelike propellers, clusters of silvery water bloom and drown the flotation gear. The element smacks the sides of the hull with a sound like a gong, and the pilot can sense this tumult in the quivering of his body. He feels the ship charging itself with power as from second to second it picks up speed. He feels the development, in these fifteen tons of matter, of a maturity that is about to make flight possible. He closes his hands over the controls, and little by little in his bare palms he receives the gift of this power. The metal organs of the controls, progressively as this gift is made him, become the messengers of the power in his hands. And when his power is ripe, then, in a gesture gentler than the culling of a flower, the pilot severs the ship from the water and establishes it in the air.

## IV

## *The Elements*

When Joseph Conrad described a typhoon he said very little about towering waves, or darkness, or the whistling of the wind in the shrouds. He knew better. Instead, he took his reader down into the hold of the vessel, packed with emigrant coolies, where the rolling and the pitching of the ship had ripped up and scattered their bags and bundles, burst open their boxes, and flung their humble belongings into a crazy heap. Family treasures painfully collected in a lifetime of poverty, pitiful mementoes so alike that nobody but their owners could have told them apart, had lost their identity and lapsed into chaos, into anonymity, into an amorphous magma. It was this human drama that Conrad described when he painted a typhoon.

Every airline pilot has flown through tornadoes, has returned out of them to the fold—to the little restaurant in Toulouse where we sat in peace under the watchful eye of the waitress—and there, recognizing his powerlessness to convey what he has been through, has given up the idea of describing hell. His descriptions, his gestures, his big words would have made the rest of us smile as if we were listening to a little boy bragging. And necessarily so. The cyclone of which I am about to speak was, physically, much the most brutal and overwhelming experience I ever underwent; and yet beyond a certain point I do not know how to convey its violence except by piling one adjective on another, so that in the end I should convey no impression at all—unless perhaps that of an embarrassing taste for exaggeration.

It took me some time to grasp the fundamental reason for this powerlessness, which is simply that I should be trying to describe a catastrophe that never took place. The reason why writers fail when they attempt to evoke horror is that horror is something invented after the fact, when one is re-creating the experience over again in the memory. Horror does not manifest itself in the world of reality. And so, in beginning my story of a revolt of the elements which I myself lived through I have no feeling that I shall write something which you will find dramatic.

I had taken off from the field at Trelew and was flying down to Comodoro-Rivadavia, in the Patagonian Argentine. Here the crust of the earth is as dented as an old boiler. The high-pressure regions over the Pacific send the winds past a gap in the Andes into a corridor fifty miles wide through which they rush to the Atlantic in a strangled and accelerated buffeting that scrapes the surface of everything in their path. The sole vegetation visible in this barren landscape is a plantation of oil derricks looking like the after-effects of a forest fire. Towering over the round hills on which the winds have left a residue of stony gravel, there rises a chain of prow-shaped, saw-toothed, razor-edged mountains stripped by the elements down to the bare rock.

For three months of the year the speed of these winds at ground level is up to a hundred miles an hour. We who flew the route knew that once we had crossed the marshes of Trelew and had reached the threshold of the zone they swept, we should recognize the winds from afar by a grey-blue tint in the atmosphere at the sight of which we would tighten our belts and shoulder-straps in preparation for what was coming. From then on we had an hour of stiff fighting and of stumbling again and again into invisible ditches of air. This was manual labor, and our muscles felt it pretty much as if we had been carrying a longshoreman's load. But it lasted only an hour. Our machines stood up under it. We had no fear of wings suddenly dropping off. Visibility was generally good, and not a problem. This section of the line was a stint, yes; it was certainly not a drama.

But on this particular day I did not like the color of the sky.

The sky was blue. Pure blue. Too pure. A hard blue sky that shone over the scraped and barren world while the fleshless vertebrae of the mountain chain flashed in the sunlight. Not a cloud. The blue sky glittered like a new-honed knife. I felt in advance the vague distaste that accompanies the prospect of physical exertion. The purity of the sky upset me. Give me a good black storm in which the enemy is plainly visible. I can measure its extent and prepare myself for its attack. I can get my hands on my adversary. But when you are flying very high in clear weather the shock of a blue storm is as disturbing as if something collapsed that had been holding up your ship in the air. It is the only time when a pilot feels that there is a gulf beneath his ship.

Another thing bothered me. I could see on a level with the

mountain peaks not a haze, not a mist, not a sandy fog, but a sort of ash-colored streamer in the sky. I did not like the look of that scarf of filings scraped off the surface of the earth and borne out to sea by the wind. I tightened my leather harness as far as it would go and I steered the ship with one hand while with the other I hung on to the longéron that ran alongside my seat. I was still flying in remarkably calm air.

Very soon came a slight tremor. As every pilot knows, there are secret little quiverings that foretell your real storm. No rolling, no pitching. No swing to speak of. The flight continues horizontal and rectilinear. But you have felt a warning drum on the wings of your plane, little intermittent rappings scarcely audible and infinitely brief, little cracklings from time to time as if there were traces of gunpowder in the air.

And then everything round me blew up.

Concerning the next couple of minutes I have nothing to say. All that I can find in my memory is a few rudimentary notions, fragments of thoughts, direct observations. I cannot compose them into a dramatic recital because there was no drama. The best I can do is to line them up in a kind of chronological order.

In the first place, I was standing still. Having banked right in order to correct a sudden drift, I saw the landscape freeze abruptly where it was and remain jiggling on the same spot. I was making no headway. My wings had ceased to nibble into the outline of the earth. I could see the earth buckle, pivot—but it stayed put. The plane was skidding as if on a toothless cogwheel.

Meanwhile I had the absurd feeling that I had exposed myself completely to the enemy. All those peaks, those crests, those teeth that were cutting into the wind and unleashing its gusts in my direction, seemed to me so many guns pointed straight at my defenseless person. I was slow to think, but the thought did come to me that I ought to give up altitude and make for one of the neighboring valleys where I might take shelter against a mountainside. As a matter of fact, whether I liked it or not I was being helplessly sucked down towards the earth.

Trapped this way in the first breaking waves of a cyclone about which I learned, twenty minutes later, that at sea level it was blowing at the fantastic rate of one hundred and fifty miles an hour, I certainly had no impression of tragedy. Now, as I write, if I shut my eyes, if I forget the plane and the flight and try to express the plain truth about what was happening to me, I find that I felt weighed down, I felt like a porter carrying a slippery load, grabbing one object in a jerky movement that sent another slithering down, so that, overcome by exasperation, the porter is tempted to let the whole load drop. There is a kind of law of the shortest distance to the image, a psychological law by which the event to which one is subjected is visualized in a symbol that represents its swiftest summing up: I was a man who, carrying a pile of plates, had slipped on a waxed floor and let his scaffolding of porcelain crash.

I found myself imprisoned in a valley. My discomfort was not less, it was greater. I grant you that a down current has never killed anybody, that the expression "flattened out by a down current" belongs to journalism and not to the language of flyers. How could air possibly pierce the ground? But here I was in a valley at the wheel of a ship that was three-quarters out of my control. Ahead of me a rocky prow swung to left and right, rose suddenly high in the air for a second like a wave over my head, and then plunged down below my horizon.

Horizon? There was no longer a horizon. I was in the wings of a theatre cluttered up with bits of scenery. Vertical, oblique, horizontal, all of plane geometry was awhirl. A hundred transversal valleys were muddled in a jumble of perspectives. Whenever I seemed about to take my bearings a new eruption would swing me round in a circle or send me tumbling wing over wing and I would have to try all over again to get clear of all this rubbish. Two ideas came into my mind. One was a discovery: for the first time I understood the cause of certain accidents in the mountains when no fog was present to explain them. For a single second, in a waltzing landscape like this, the flyer had been unable to distinguish between vertical mountainsides and horizontal planes. The other idea was a fixation: The sea is flat: I shall not hook anything out at sea.

I banked—or should I use that word to indicate a vague and stubborn jockeying through the east-west valleys? Still nothing pathetic to report. I was wrestling with chaos, was wearing myself out in a battle with chaos, struggling to keep in the air a gigantic house of cards that kept collapsing despite all I could do. Scarcely the faintest twinge of fear went through me when one of the walls of my prison rose suddenly like a tidal wave over my head. My heart hardly skipped a beat when I was tripped up by one of the whirling eddies of air that the sharp ridge darted into my ship. If I felt anything unmistakably in the haze of confused feelings and notions that came over me each time one of these powder magazines blew up, it was a feeling of respect. I respected that sharp-toothed ridge. I respected that peak. I respected that dome. I respected that transversal valley opening out into my valley and about to toss me God knew how violently as soon as its torrent of wind flowed into the one on which I was being borne along.

What I was struggling against, I discovered, was not the wind but the ridge itself, the crest, the rocky peak. Despite my distance from it, it was the wall of rock I was fighting with. By some trick of invisible prolongation, by the play of a secret set of muscles, this was what was pummeling me. It was against this that I was butting my head. Before me on the right I recognized the peak of Salamanca, a perfect cone which, I knew, dominated the sea. It cheered me to think I was about to escape out to sea. But first I should have to wrestle with the gale off that peak, try to avoid its down-crushing blow. The peak of Salamanca was a giant. I was filled with respect for the peak of Salamanca.

There had been granted me one second of respite. Two seconds. Something was collecting itself into a knot, coiling itself up, growing taut. I sat amazed. I opened astonished eyes. My whole plane seemed to be shivering, spreading outward, swelling up. Horizontal and stationary it was, yet lifted before I knew it fifteen hundred feet straight into the air in a kind of apotheosis. I who for forty minutes had not been able to climb higher than two hundred feet off the ground was suddenly able to look down on the enemy. The plane quivered as if in boiling water. I could see the wide waters of the ocean. The valley opened out into this ocean, this salvation.—And at that very moment, without any warning whatever, half a mile from Salamanca, I was suddenly struck straight in the midriff by the gale off that peak and sent hurtling out to sea.

There I was, throttle wide open, facing the coast. At right angles to the coast and facing it. A lot had happened in a single minute. In the first place, I had not flown out to sea. I had been spat out to sea by a monstrous cough, vomited out of my valley as from the mouth of a howitzer. When, what seemed to me instantly, I banked in order to put myself where I wanted to be in respect of the coast-line, I saw that the coast-line was a mere blur, a characterless strip of blue; and I was five miles out to sea. The mountain range stood up like a crenelated fortress against the pure sky while the cyclone crushed me down to the surface of the waters. How hard that wind was blowing I found out as soon as I tried to climb, as soon as I became conscious of my disastrous mistake: throttle wide open, engines running at my maximum, which was one hundred and fifty miles an hour, my plane hanging sixty feet over the water, I was unable to budge. When a wind like this one attacks a tropical forest it swirls through the branches like a flame, twists them into corkscrews, and uproots giant trees as if they were radishes. Here, bounding off the mountain range, it was leveling out to sea.

Hanging on with all the power in my engines, face to the coast, face to that wind where each gap in the teeth of the range sent forth a stream of air like a long reptile, I felt as if I were clinging to the tip of a monstrous whip that was cracking over the sea.

In this latitude the South American continent is narrow and the Andes are not far from the Atlantic. I was struggling not merely against the whirling winds that blew off the east-coast range, but more likely also against a whole sky blown down upon me off the peaks of the Andean chain. For the first time in four years of airline flying I began to worry about the strength of my wings. Also, I was fearful of bumping the sea—not because of the down currents which, at sea level, would necessarily provide me with a horizontal air mattress, but because of the helplessly acrobatic positions in which this wind was buffeting me. Each time that I was tossed I became afraid that I might be unable to straighten out. Besides, there was a chance that I should find myself out of fuel and simply drown. I kept expecting the gasoline pumps to stop priming, and indeed the plane was so violently shaken up that in the half-filled tanks as well as in the gas lines the gasoline was sloshing round, not coming through, and the engines, instead of their steady roar, were sputtering in a sort of dot-and-dash series of uncertain growls.

I hung on, meanwhile, to the controls of my heavy transport plane, my attention monopolized by the physical struggle and my mind occupied by the very simplest thoughts. I was feeling practically nothing as I stared down at the imprint made by the wind on the sea. I saw a series of great white puddles, each perhaps eight hundred yards in extent. They were running towards me at a speed of one hundred and fifty miles an hour where the down-surging windspouts broke against the surface of the sea in a succession of horizontal explosions. The sea was white and it was green—white with the whiteness of crushed sugar and green in puddles the color of emeralds. In this tumult one wave was indistinguishable from another. Torrents of air were pouring down upon the sea. The winds were sweeping past in giant gusts as when, before the autumn harvests, they blow a great flowing change of color over a wheatfield. Now and again the water went incongruously transparent between the white pools, and I could see a green and black sea-bottom. And then the great glass of the sea would be shattered anew into a thousand glittering fragments.

It seemed hopeless. In twenty minutes of struggle I had not moved forward a hundred yards. What was more, with flying as hard as it was out here five miles from the coast, I wondered how I could possibly buck the winds along the shore, assuming I was able to fight my way in. I was a perfect target for the enemy there on shore. Fear, however, was out of the question. I was incapable of thinking. I was emptied of everything except the vision of a very simple act. I must straighten out. Straighten out. Straighten out.

There were moments of respite, nevertheless. I dare say those moments themselves were equal to the worst storms I had hitherto met, but by comparison with the cyclone they were moments of relaxation. The urgency of fighting off the wind was not quite so great. And I could tell when these intervals were coming. It was not I who moved towards those zones of relative calm, those almost green oases clearly painted on the sea, but they that flowed towards me. I could read clearly in the waters the advertisement of a habitable province. And with each interval of repose the power to feel and to think was restored to me. Then, in those moments, I began to feel I was doomed. Then was the time that little by little I began to tremble for myself. So much so that each time I saw the unfurling of a new wave of the white offensive I was seized by a brief spasm of panic which lasted until the exact instant when, on the edge of that bubbling cauldron, I bumped into the invisible wall of wind. That restored me to numbness again.

Up! I wanted to be higher up. The next time I saw one of those green zones of calm it seemed to me deeper than before and I began to be hopeful of getting out. If I could climb high enough, I thought, I would find other currents in which I could make some headway. I took advantage of the truce to essay a swift climb. It was hard. The enemy had not weakened. Three hundred feet. Six hundred feet. If I could get up to three thousand feet I was safe, I said to myself. But there on the horizon I saw again that white pack unleashed in my direction. I gave it up. I did not want them at my throat again; I did not want to be caught off balance. But it was too late. The first blow sent me rolling over and over and the sky became a slippery dome on which I could not find a footing.

One has a pair of hands and they obey. How are one's orders transmitted to one's hands?

I had made a discovery that horrified me: my hands were numb. My hands were dead. They sent me no message. Probably they had been numb a long time and I had not noticed it. The pity was that I had noticed it, had raised the question. That was serious.

Lashed by the wind, the wings of the plane had been dragging and jerking at the cables by which they were controlled from the wheel, and the wheel in my hands had not ceased jerking a single second. I had been gripping the wheel with all my might for forty minutes, fearful lest the strain snap the cables. So desperate had been my grip that now I could not feel my hands.

What a discovery! My hands were not my own. I looked at them and decided to lift a finger: it obeyed me. I looked away and issued the same order: now I could not feel whether the finger had obeyed or not. No message had reached me. I thought: "Suppose my hands were to open: how would I know it?" I swung my head round and looked again: my hands were still locked round the wheel. Nevertheless, I was afraid. How can a man tell the difference between the sight of a hand opening and the decision to open that hand, when there is no longer an exchange of sensations between the hand and the brain? How can one tell the difference between an image and an act of the will? Better stop thinking of the picture of open hands. Hands live a life of their own. Better not offer them this monstrous temptation. And I began to chant a silly litany which went on uninterruptedly until this flight was over. A single thought. A single image. A single phrase tirelessly chanted over and over again: "I shut my hands. I shut my hands. I shut my hands." All of me was condensed into that phrase and for me the white sea, the whirling eddies, the saw-toothed range ceased to exist. There was only "I shut my hands." There was no danger, no cyclone, no land unattained. Somewhere there was a pair of rubber hands which, once they let go the wheel, could not possibly come alive in time to recover from the tumbling drop into the sea.

I had no thoughts. I had no feelings except the feeling of being emptied out. My strength was draining out of me and so was my impulse to go on fighting. The engines continued their dot-and-dash sputterings, their little crashing noises that were like the intermittent cracklings of a ripping canvas. Whenever they were silent longer than a second I felt as if a heart had stopped beating. There! that's the end. No, they've started up again.

The thermometer on the wing, I happened to see, stood at twenty below zero, but I was bathed in sweat from head to foot. My face was running with perspiration. What a dance! Later I was to discover that my storage batteries had been jerked out of their steel flanges and hurtled up through the roof of the plane. I did not know then, either, that the ribs on my wings had come unglued and that certain of my steel cables had been sawed down to the last thread. And I continued to feel strength and will oozing out of me. Any minute now I should be overcome by the indifference born of utter weariness and by the mortal yearning to take my rest.

What can I say about this? Nothing. My shoulders ached. Very painfully. As if I had been carrying too many sacks too heavy for me. I leaned forward. Through a green transparency I saw sea-bottom so close that I could make out all the details. Then the wind's hand brushed the picture away.

In an hour and twenty minutes I had succeeded in climbing to nine hundred feet. A little to the south—that is, on my left—I could see a long trail on the surface of the sea, a sort of blue stream. I decided to let myself drift as far down as that stream. Here where I was, facing west, I was as good as motionless, unable either to advance or retreat. If I could reach that blue pathway, which must be lying in the shelter of something not the cyclone, I might be able to move in slowly to the coast. So I let myself drift to the left. I had the feeling, meanwhile, that the wind's violence had perhaps slackened.

It took me an hour to cover the five miles to shore. There in the shelter of a long cliff I was able to finish my journey south. Thereafter I succeeded in keeping enough altitude to fly inland to the field that was my destination. I was able to stay up at nine hundred feet. It was very stormy, but nothing like the cyclone I had come out of. That was over.

On the ground I saw a platoon of soldiers. They had been sent down to watch for me. I landed near by and we were a whole hour getting the plane into the hangar. I climbed out of the cockpit and walked off. There was nothing to say. I was very sleepy. I kept moving my fingers, but they stayed numb. I could not collect my thoughts enough to decide whether or not I had been afraid. Had I been afraid? I couldn't say. I had witnessed a strange sight. What strange sight? I couldn't say. The sky was blue and the sea was white. I felt I ought to tell someone about it since I was back from so far away! But I had no grip on what I had been through. "Imagine a white sea . . . very white . . . whiter still." You cannot convey things to people by piling up adjectives, by stammering.

You cannot convey anything because there is nothing to convey. My shoulders were aching. My insides felt as if they had been crushed in by a terrible weight. You cannot make drama out of that, or out of the cone-shaped peak of Salamanca. That peak was charged like a powder magazine; but if I said so people would laugh. I would myself. I respected the peak of Salamanca. That is my story. And it is not a story.

There is nothing dramatic in the world, nothing pathetic, except in human relations. The day after I landed I might get emotional, might dress up my adventure by imagining that I who was alive and walking on earth was living through the hell of a cyclone. But that would be cheating, for the man who fought tooth and nail against that cyclone had nothing in common with the fortunate man alive the next day. He was far too busy.

I came away with very little booty indeed, with no more than this meagre discovery, this contribution: How can one tell an act of the will from a simple image when there is no transmission of sensation?

I could perhaps succeed in upsetting you if I told you some story of a child unjustly punished. As it is, I have involved you in a cyclone, probably without upsetting you in the least. This is no novel experience for any of us. Every week men sit comfortably at the cinema and look on at the bombardment of some Shanghai or other, some Guernica, and marvel without a trace of horror at the long fringes of ash and soot that twist their slow way into the sky from those man-made volcanoes. Yet we all know that together with the grain in the granaries, with the heritage of generations of men, with the treasures of families, it is the burning flesh of children and their elders that, dissipated in smoke, is slowly fertilizing those black cumuli.

The physical drama itself cannot touch us until some one points out its spiritual sense.

# V

## *The Plane and the Planet*

The airplane has unveiled for us the true face of the earth. For centuries, highways had been deceiving us. We were like that queen who determined to move among her subjects so that she might learn for herself whether or not they rejoiced in her reign. Her courtiers took advantage of her innocence to garland the road she traveled and set dancers in her path. Led forward on their halter, she saw nothing of her kingdom and could not know that over the countryside the famished were cursing her.

Even so have we been making our way along the winding roads. Roads avoid the barren lands, the rocks, the sands. They shape themselves to man's needs and run from stream to stream. They lead the farmer from his barns to his wheatfields, receive at the thresholds of stables the sleepy cattle and pour them forth at dawn into meadows of alfalfa. They join village to village, for between villages marriages are made.

And even when a road hazards its way over the desert, you will see it make a thousand détours to take its pleasure at the oases. Thus, led astray by the divagations of roads, as by other indulgent fictions, having in the course of our travels skirted so many well-watered lands, so many orchards, so many meadows, we have from the beginning of time embellished the picture of our prison. We have elected to believe that our planet was merciful and fruitful.

But a cruel light has blazed, and our sight has been sharpened. The plane has taught us to travel as the crow flies. Scarcely have we taken off when we abandon these winding highways that slope down to watering troughs and stables or run away to towns dreaming in the shade of their trees. Freed henceforth from this happy servitude, delivered from the need of fountains, we set our course for distant destinations. And then, only, from the height of our rectilinear trajectories, do we discover the essential foundation, the fundament of rock and sand and salt in which here and there and from time to time life like a little moss in the crevices of ruins has risked its precarious existence.

We to whom humble journeyings were once permitted have now been transformed into physicists, biologists, students of the civilizations that beautify the depths of valleys and now and again, by some miracle, bloom like gardens where the climate allows. We are able to judge man in cosmic terms, scrutinize him through our portholes as through instruments of the laboratory. I remember a few of these scenes.

## I

The pilot flying towards the Straits of Magellan sees below him, a little to the south of the Gallegos River, an ancient lava flow, an erupted waste of a thickness of sixty feet that crushes down the plain on which it has congealed. Farther south he meets a second flow, then a third; and thereafter every hump on the globe, every mound a few hundred feet high, carries a crater in its flank. No Vesuvius rises up to reign in the clouds; merely, flat on the plain, a succession of gaping howitzer mouths.

This day, as I fly, the lava world is calm. There is something surprising in the tranquillity of this deserted landscape where once a thousand volcanoes boomed to each other in their great subterranean organs and spat forth their fire. I fly over a world mute and abandoned, strewn with black glaciers.

South of these glaciers there are yet older volcanoes veiled with the passing of time in a golden sward. Here and there a tree rises out of a crevice like a plant out of a cracked pot. In the soft and yellow light the plain appears as luxuriant as a garden; the short grass seems to civilize it, and round its giant throats there is scarcely a swelling to be seen. A hare scampers off; a bird wheels in the air; life has taken possession of a new planet where the decent loam of our earth has at last spread over the surface of the star.

Finally, crossing the line into Chile, a little north of Punta Arenas, you come to the last of the craters, and here the mouths have been stopped with earth. A silky turf lies snug over the curves of the volcanoes, and all is suavity in the scene. Each fissure in the crust is sutured up by this tender flax. The earth is smooth, the slopes are gentle; one forgets the travail that gave them birth. This turf effaces from the flanks of the hillocks the sombre sign of their origin.

We have reached the most southerly habitation of the world, a town born of the chance presence of a little mud between the timeless lava and the austral ice. So near the black scoria, how thrilling it is to feel the miraculous nature of man! What a strange encounter! Who knows how, or why, man visits these gardens ready to hand, habitable for so short a time—a geologic age—for a single day blessed among days?

I landed in the peace of evening. Punta Arenas! I leaned against a fountain and looked at the girls in the square. Standing there within a couple of feet of their grace, I felt more poignantly than ever the human mystery.

In a world in which life so perfectly responds to life, where flowers mingle with flowers in the wind's eye, where the swan is the familiar of all swans, man alone builds his isolation. What a space between men their spiritual natures create! A girl's reverie isolates her from me, and how shall I enter into it? What can one know of a girl who passes, walking with slow steps homeward, eyes lowered, smiling to herself, filled with adorable inventions and with fables? Out of the thoughts, the voice, the silences of a lover, she can form an empire, and thereafter she sees in all the world but him a people of barbarians. More surely than if she were on another planet, I feel her to be locked up in her language, in her secret, in her habits, in the singing echoes of her memory. Born yesterday of the volcanoes, of greenswards, of brine of the sea, she walks here already half divine.

Punta Arenas! I lean against a fountain. Old women come up to draw water: of their drama I shall know nothing but these gestures of farm servants. A child, his head against a wall, weeps in silence: there will remain of him in my memory only a beautiful child forever inconsolable. I am a stranger. I know nothing. I do not enter into their empires. Man in the presence of man is as solitary as in the face of a wide winter sky in which there sweeps, never to be tamed, a flight of trumpeting geese.

How shallow is the stage on which this vast drama of human hates and joys and friendships is played! Whence do men draw this passion for eternity, flung by chance as they are upon a scarcely cooled bed of lava, threatened from the beginning by the deserts that are to be, and under the constant menace of the snows? Their civilizations are but fragile gildings: a volcano can blot them out, a new sea, a sand-storm.

This town seemed to be built upon a true humus, a soil one might imagine to be as rich as the wheatlands of the Beauce. These men live heedless of the fact that, here as elsewhere, life is a luxury; and that nowhere on the globe is the soil really rich beneath the feet of men.

Yet, ten miles from Punta Arenas there is a lake that ought to be reminding them of this. Surrounded by stunted trees and squat huts, as modest as a pool in a farm-yard, this lake is subject to the preternatural pull of the tides. Night and day, among the peaceful realities of swaying reeds and playing children, it performs its slow respiration, obedient to unearthly laws. Beneath the glassy surface, beneath the motionless ice, beneath the keel of the single dilapidated bark on the waters, the energy of the moon is at work. Ocean eddies stir in the depths of this black mass. Strange digestions take their peristaltic course there and down as far as the Straits of Magellan, under the thin layer of grasses and flowers. This lake that is a hundred yards wide, that laps the threshold of a town which seems to be built on man's own earth and where men believe themselves secure, beats with the pulse of the sea.

## II

But by the grace of the airplane I have known a more extraordinary experience than this, and have been made to ponder with even more bewilderment the fact that this earth that is our home is yet in truth a wandering star.

A minor accident had forced me down in the Rio de Oro region, in Spanish Africa. Landing on one of those table-lands of the Sahara which fall away steeply at the sides, I found myself on the flat top of the frustrum of a cone, an isolated vestige of a plateau that had crumbled round the edges. In this part of the Sahara such truncated cones are visible from the air every hundred miles or so, their smooth surfaces always at about the same altitude above the desert and their geologic substance always identical. The surface sand is composed of minute and distinct shells; but progressively as you dig along a vertical section, the shells become more fragmentary, tend to cohere, and at the base of the cone form a pure calcareous deposit.

Without question, I was the first human being ever to wander over this . . . this iceberg: its sides were remarkably steep, no Arab could have climbed them, and no European had as yet ventured into this wild region.

I was thrilled by the virginity of a soil which no step of man or beast had sullied. I lingered there, startled by this silence that never had been broken. The first star began to shine, and I said to myself that this pure surface had lain here thousands of years in sight only of the stars.

But suddenly my musings on this white sheet and these shining stars were endowed with a singular significance. I had kicked against a hard, black stone, the size of a man's fist, a sort of moulded rock of lava incredibly present on the surface of a bed of shells a thousand feet deep. A sheet spread beneath an apple-tree can receive only apples; a sheet spread beneath the stars can receive only star-dust. Never had a stone fallen from the skies made known its origin so unmistakably.

And very naturally, raising my eyes, I said to myself that from the height of this celestial apple-tree there must have dropped other fruits, and that I should find them exactly where they fell, since never from the beginning of time had anything been present to displace them.

Excited by my adventure, I picked up one and then a second and then a third of these stones, finding them at about the rate of one stone to the acre. And here is where my adventure became magical, for in a striking foreshortening of time that embraced thousands of years. I had become the witness of this miserly rain from the stars. The marvel of marvels was that there on the rounded back of the planet, between this magnetic sheet and those stars, a human consciousness was present in which as in a mirror that rain could be reflected.

## III

Once, in this same mineral Sahara, I was taught that a dream might partake of the miraculous. Again I had been forced down, and until day dawned I was helpless. Hillocks of sand offered up their luminous slopes to the moon, and blocks of shadow rose to share the sands with the light. Over the deserted work-yard of darkness and moonray there reigned a peace as of work suspended and a silence like a trap, in which I fell asleep.

When I opened my eyes I saw nothing but the pool of nocturnal sky, for I was lying on my back with outstretched arms, face to face with that hatchery of stars. Only half awake, still unaware that those depths were sky, having no roof between those depths and me, no branches to screen them, no root to cling to, I was seized with vertigo and felt myself as if flung forth and plunging downward like a diver.

But I did not fall. From nape to heel I discovered myself bound to earth. I felt a sort of appeasement in surrendering to it my weight. Gravitation had become as sovereign as love. The earth, I felt, was supporting my back, sustaining me, lifting me up, transporting me through the immense void of night. I was glued to our planet by a pressure like that with which one is glued to the side of a car on a curve. I leaned with joy against this admirable breastwork, this solidity, this security, feeling against my body this curving bridge of my ship.

So convinced was I that I was in motion, that I should have heard without astonishment, rising from below, a creaking of something material adjusting itself to the effort, that groaning of old sailing vessels as they heel, that long sharp cry drawn from pinnaces complaining of their handling. But silence continued in the layers of the earth, and this density that I could feel at my shoulders continued harmonious, sustained, unaltered through eternity. I was as much the inhabitant of this homeland as the bodies of dead galley-slaves, weighted with lead, were the inhabitants of the sea.

I lay there pondering my situation, lost in the desert and in danger, naked between sky and sand, withdrawn by too much silence from the poles of my life. I knew that I should wear out days and weeks returning to them if I were not sighted by some plane, or if next day the Moors did not find and murder me. Here I possessed nothing in the world. I was no more than a mortal strayed between sand and stars, conscious of the single blessing of breathing. And yet I discovered myself filled with dreams.

They came to me soundlessly, like the waters of a spring, and in the beginning I could not understand the sweetness that was invading me. There was neither voice nor vision, but the presentiment of a presence, of a warmth very close and already half guessed. Then I began to grasp what was going on, and shutting my eyes I gave myself up to the enchantments of my memory.

Somewhere there was a park dark with firs and linden-trees and an old house that I loved. It mattered little that it was far away, that it could not warm me in my flesh, nor shelter me, reduced here to the rôle of dream. It was enough that it existed to fill my night with its presence. I was no longer this body flung up on a stand; I oriented myself; I was the child of this house, filled with the memory of its odors, with the cool breath of its vestibules, with the voices that had animated it, even to the very frogs in the pools that came here to be with me. I needed these thousand landmarks to identify myself, to discover of what absences the savor of this desert was composed, to find a meaning in this silence made of a thousand silences, where the very frogs were silent.

No, I was no longer lodged between sand and stars. I was no longer receiving from this scene its chill message. And I had found out at last the origin of the feeling of eternity that came over me in this wilderness. I had been wrong to believe it was part of sky and sand. I saw again the great stately cupboards of our house. Their doors opened to display piles of linen as white as snow. They opened on frozen stores of snow. The old housekeeper trotted like a rat from one cupboard to the next, forever counting, folding, unfolding, recounting the white linen; exclaiming, "Oh, good Heavens, how terrible!" at each sign of wear which threatened the eternity of the house; running instantly to burn out her eyes under a lamp so that the woof of these altar cloths should be repaired, these three-master's sails be mended, in the service of something greater than herself—a god, a ship.

Ah, I owe you a page, Mademoiselle! When I came home from my first journeyings I found you needle in hand, up to the knees in your white surplices, each year a little more wrinkled, a little more round-shouldered, still preparing for our slumbers those sheets without creases, for our dinners those cloths without seams, those feasts of crystal and of snow.

I would go up to see you in your sewing-room, would sit down beside you and tell you of the dangers I had run in order that I might thrill you, open your eyes to the world, corrupt you. You would say that I hadn't changed a whit. Already as a child I had torn my shirts—"How terrible!"—and skinned my knees, coming home as day fell to be bandaged.

No, Mademoiselle, no! I have not come back from the other end of the park but from the other end of the world! I have brought back with me the acrid smell of solitude, the tumult of sandstorms, the blazing moonlight of the tropics! "Of course!" you would say. "Boys *will* run about, break their bones and think themselves great fellows."

No, Mademoiselle, no! I have seen a good deal more than the

shadows in our park. If you knew how insignificant these shadows are, how little they mean beside the sands, the granite, the virgin forests, the vast swamplands of the earth! Do you realize that there are lands on the globe where, when men meet you, they bring up their rifles to their cheeks? Do you know that there are deserts on earth where men lie down on freezing nights to sleep without roof or bed or snowy sheet? "What a wild lad!" you would say.

I could no more shake her faith than I could have shaken the faith of a candle-woman in a church. I pitied her humble destiny which had made her blind and deaf.

But that night in the Sahara, naked between the stars and the sand, I did her justice.

What is going on inside me I cannot tell. In the sky a thousand stars are magnetized, and I lie glued by the swing of the planet to the sand. A different weight brings me back to myself. I feel the weight of my body drawing me towards so many things. My dreams are more real than these dunes, than that moon, than these presences. My civilization is an empire more imperious than this empire. The marvel of a house is not that it shelters or warms a man, nor that its walls belong to him. It is that it leaves its trace on the language. Let it remain a sign. Let it form, deep in the heart, that obscure range from which, as waters from a spring, are born our dreams.

Sahara, my Sahara! You have been bewitched by an old woman at a spinning-wheel!

# VI

## *Oasis*

I have already said so much about the desert that before speaking of it again I should like to describe an oasis. The oasis that comes into my mind is not, however, remote in the deep Sahara. One of the miracles of the airplane is that it plunges a man directly into the heart of mystery. You are a biologist studying, through your porthole, the human ant-hill, scrutinizing objectively those towns seated in their plain at the centre of their highways which go off like the spokes of a wheel and, like arteries, nourish them with the quintessence of the fields. A needle trembles on your manometer, and this green clump below you becomes a universe. You are the prisoner of a greensward in a slumbering park.

Space is not the measure of distance. A garden wall at home may enclose more secrets than the Great Wall of China, and the soul of a little girl is better guarded by silence than the Sahara's oases by the surrounding sands. I dropped down to earth once somewhere in the world. It was near Concordia, in the Argentine, but it might have been anywhere at all, for mystery is everywhere.

A minor mishap had forced me down in a field, and I was far from dreaming that I was about to live through a fairy-tale. The old Ford in which I was driven to town betokened nothing extraordinary, and the same was to be said for the unremarkable couple who took me in.

"We shall be glad to put you up for the night," they said.

But round a corner of the road, in the moonlight, I saw a clump of trees, and behind those trees a house. What a queer house! Squat, massive, almost a citadel guarding behind its tons of stone

I knew not what treasure. From the very threshold this legendary castle promised an asylum as assured, as peaceful, as secret as a monastery.

Then two young girls appeared. They seemed astonished to see me, examined me gravely as if they had been two judges posted on the confines of a forbidden kingdom, and while the younger of them sulked and tapped the ground with a green switch, they were introduced:

"Our daughters."

The girls shook hands without a word but with a curious air of defiance, and disappeared. I was amused and I was charmed. It was all as simple and silent and furtive as the first word of a secret.

"The girls are shy," their father said, and we went into the house.

One thing that I had loved in Paraguay was the ironic grass that showed the tip of its nose between the pavements of the capital, that slipped in on behalf of the invisible but ever-present virgin forest to see if man still held the town, if the hour had not come to send all these stones tumbling.

I liked the particular kind of dilapidation which in Paraguay was the expression of an excess of wealth. But here, in Concordia, I was filled with wonder. Here everything was in a state of decay, but adorably so, like an old oak covered with moss and split in places with age, like a wooden bench on which generations of lovers had come to sit and which had grown sacred. The wainscoting was worn, the hinges rusted, the chairs rickety. And yet, though nothing had ever been repaired, everything had been scoured with zeal. Everything was clean, waxed, gleaming.

The drawing-room had about it something extraordinarily intense, like the face of a wrinkled old lady. The walls were cracked, the ceiling stripped; and most bewildering of all in this bewildering house was the floor: it had simply caved in. Waxed, varnished and polished though it was, it swayed like a ship's gangway. A strange house, evoking no neglect, no slackness, but rather an extraordinary respect. Each passing year had added something to its charm, to the complexity of its visage and its friendly atmosphere, as well as to the dangers encountered on the journey from the drawing-room to the dining-room.

"Careful!"

There was a hole in the floor; and I was warned that if I stepped into it I might easily break a leg. This was said as simply as "Don't stroke the dog, he bites." Nobody was responsible for the hole, it was the work of time. There was something lordly about this sovereign contempt for apologies.

Nobody said, "We could have these holes repaired; we are well enough off; but . . ." And neither did they say—which was true enough—"we have taken this house from the town under a thirty-year lease. They should look after the repairs. But they won't, and we won't, so . . ." They disdained explanation, and this superiority to circumstance enchanted me. The most that was said was:

"The house is a little run down, you see."

Even this was said with such an air of satisfaction that I suspected my friends of not being saddened by the fact. Do you see a crew of bricklayers, carpenters, cabinet-workers, plasterers intruding their sacrilegious tools into so vivid a past, turning this in a week into a house you would never recognize, in which the family would feel that they were visiting strangers? A house without secrets, without recesses, without mysteries, without traps beneath the feet, or dungeons, a sort of town-hall reception room?

In a house with so many secret passages it was natural that the daughters should vanish before one's eyes. What must the attics be, when the drawing-room already contained all the wealth of an attic? When one could guess already that, the least cupboard opened, there would pour out sheaves of yellowed letters, grandpapa's receipted bills, more keys than there were locks and not one of which of course would fit any lock. Marvelously useless keys that confounded the reason and made it muse upon subterranean chambers, buried chests, treasures.

"Shall we go in to dinner?"

We went in to dinner. Moving from one room to the next I inhaled in passing that incense of an old library which is worth all the perfumes of the world. And particularly I liked the lamps being carried with us. Real lamps, heavy lamps, transported from room to room as in the time of my earliest childhood; stirring into motion as they passed great wondrous shadows on the walls. To pick one up was to displace bouquets of light and great black palms. Then, the lamps finally set down, there was a settling into motionlessness of the beaches of clarity and the vast reserves of surrounding darkness in which the wainscoting went on creaking.

As mysteriously and as silently as they had vanished, the girls reappeared. Gravely they took their places. Doubtless they had fed their dogs, their birds; had opened their windows on the bright night and breathed in the smell of the woods brought by the night wind. Now, unfolding their napkins, they were inspecting me cautiously out of the corners of their eyes, wondering whether or not they were going to make place for me among their domestic animals. For among others they had an iguana, a mongoose, a fox, a monkey, and bees. All these lived promiscuously together without quarreling in this new earthly paradise. The girls reigned over all the animals of creation, charming them with their little hands, feeding them, watering them, and telling them tales to which all, from mongoose to bees, gave ear.

I firmly expected that these alert young girls would employ all their critical faculty, all their shrewdness, in a swift, secret, and irrevocable judgment upon the male who sat opposite them.

When I was a child my sisters had a way of giving marks to guests who were honoring our table for the first time. Conversation might languish for a moment, and then in the silence we would hear the sudden impact of "Sixty!"—a word that could tickle only the family, who knew that one hundred was par. Branded by this low mark, the guest would all unknowing continue to spend himself in little courtesies while we sat screaming inwardly with delight.

Remembering that little game, I was worried. And it upset me a bit more to feel my judges so keen. Judges who knew how to distinguish between candid animals and animals that cheated; who could tell from the tracks of the fox whether he was in a good temper or not; whose instinct for inner movements was so sure and deep.

I liked the sharp eyes of these straightforward little souls, but I should so much have preferred that they play some other game. And yet, in my cowardly fear of their "sixty" I passed them the salt, poured out their wine; though each time that I raised my eyes I saw in their faces the gentle gravity of judges who were not to be bought.

Flattery itself was useless: they knew no vanity. Although they knew not it, they knew a marvelous pride, and without any help from me they thought more good of themselves than I should have dared utter. It did not even occur to me to draw any prestige from my craft, for it is extremely dangerous to clamber up to the topmost branches of a plane-tree simply to see if the nestlings are doing well or to say good morning to one's friends.

My taciturn young friends continued their inspection so imperturbably, I met so often their fleeting glances, that soon I stopped talking. Silence fell, and in that silence I heard something hiss faintly under the floor, rustle under the table, and then stop. I raised a pair of puzzled eyes. Thereupon, satisfied with her examination but applying her last touchstone, as she bit with savage young teeth into her bread the younger daughter explained to me with a candor by which she hoped to slaughter the barbarian (if that was what I was):

"It's the snakes."

And content, she said no more, as if that explanation should have sufficed for anyone in whom there remained a last glimmer of intelligence. Her sister sent a lightning glance to spy out my immediate reflex, and both bent with the gentlest and most ingenuous faces in the world over their plates.

"Ah! Snakes, are they?"

Naturally the words escaped from me against my will. This that had been gliding between my legs, had been brushing my calves, was snakes!

Fortunately for me, I smiled. Effortlessly. They would have known if it had been otherwise. I smiled because my heart was light, because each moment this house was more and more to my liking. And also because I wanted to know more about the snakes. The elder daughter came to my rescue.

"They nest in a hole under the table."

And her sister added: "They go back into their nest at about ten o'clock. During the day they hunt."

Now it was my turn to look at them out of the corner of the eye. What shrewdness! what silent laughter behind those candid faces! And what sovereignty they exercised, these princesses guarded by snakes! Princesses for whom there existed no scorpion, no wasp, no serpent, but only little souls of animals!

As I write, I dream. All this is very far away. What has become of these two fairy princesses? Girls so fine-grained, so upright, have certainly attracted husbands. Have they changed, I wonder? What do they do in their new houses? Do they feel differently now about the jungle growth and the snakes? They had been fused with something universal, and then the day had come when the woman had awakened in the maiden, when there had surged in her a longing to find someone who deserved a "Ninety-five." The dream of a ninety-five is a weight on the heart.

And then an imbecile had come along. For the first time those sharp eyes were mistaken and they dressed him in gay colors. If the imbecile recited verse he was thought a poet. Surely he must understand the holes in the floor, must love the mongoose! The trust one put in him, the swaying of the snakes between his legs under the table—surely this must flatter him! And that heart which was a wild garden was given to him who loved only trim lawns. And the imbecile carried away the princess into slavery.

# VII

## *Men of the Desert*

These, then, were some of the treasures that passed us by when for weeks and months and years we, pilots of the Sahara line, were prisoners of the sands, navigating from one stockade to the next with never an excursion outside the zone of silence. Oases like these did not prosper in the desert; these memories it dismissed as belonging to the domain of legend. No doubt there did gleam in distant places scattered round the world—places to which we should return once our work was done—there did gleam lighted windows. No doubt somewhere there did sit young girls among their white lemurs or their books, patiently compounding souls as rich in delight as secret gardens. No doubt there did exist such creatures waxing in beauty. But solitude cultivates a strange mood.

I know that mood. Three years of the desert taught it to me. Something in one's heart takes fright, not at the thought of growing old, not at feeling one's youth used up in this mineral universe, but at the thought that far away the whole world is ageing. The trees have brought forth their fruit; the grain has ripened in the fields; the women have bloomed in their loveliness. But the season is advancing and one must make haste; but the season is advancing and still one cannot leave; but the season is advancing . . . and other men will glean the harvest.

Many a night have I savored this taste of the irreparable, wandering in a circle round the fort, our prison, under the burden of the trade-winds. Sometimes, worn out by a day of flight, drenched in the humidity of the tropical climate, I have felt my heart beat in me like the wheels of an express train; and suddenly, more immediately than when flying, I have felt myself on a journey. A journey through time. Time was running through my fingers like the fine sand of the dunes; the poundings of my heart were bearing me onward towards an unknown future.

Ah, those fevers at night after a day of work in the silence! We seemed to ourselves to be burning up, like flares set out in the solitude.

And yet we knew joys we could not possibly have known elsewhere. I shall never be able to express clearly whence comes this pleasure men take from aridity, but always and everywhere I have seen men attach themselves more stubbornly to barren lands than to any other. Men will die for a calcined, leafless, stony mountain. The nomads will defend to the death their great store of sand as if it were a treasure of gold dust. And we, my comrades and I, we too have loved the desert to the point of feeling that it was there we had lived the best years of our lives. I shall describe for you our stations (Port Etienne, Villa Cisneros, Cape Juby, were some of their names) and shall narrate for you a few of our days.

## I

I succumbed to the desert as soon as I saw it, and I saw it almost as soon as I had won my wings. As early as the year 1926 I was transferred out of Europe to the Dakar-Juby division, where the Sahara meets the Atlantic and where, only recently, the Arabs had murdered two of our pilots, Erable and Gourp. In those days our planes frequently fell apart in mid-air, and because of this the African divisions were always flown by two ships, one without the mails trailing and convoying the other, prepared to take over the sacks in the event the mail plane broke down.

Under orders, I flew an empty ship down to Agadir. From Agadir I was flown to Dakar as a passenger, and it was on that flight that the vast sandy void and the mystery with which my imagination could not but endow it first thrilled me. But the heat was so intense that despite my excitement I dozed off soon after we left Port Etienne. Riguelle, who was flying me down, moved out to sea a couple of miles in order to get away from the sizzling surface of sand. I woke up, saw in the distance the thin white line of the coast, and said to myself fearfully that if anything went wrong we should surely drown. Then I dozed off again.

I was startled out of my sleep by a crash, a sudden silence, and then the voice of Riguelle saying, "Damn! There goes a connecting rod!" As I half rose out of my seat to send a regretful look at that white coast-line, now more precious than ever, he shouted to me angrily to stay as I was. I knew Riguelle had been wrong to go out to sea; I had been on the point of mentioning it; and now I felt a complete and savage satisfaction in our predicament. "This," I said to myself, "will teach him a lesson."

But this gratifying sense of superiority could obviously not last very long. Riguelle sent the plane earthward in a long diagonal line that brought us within sixty feet of the sand—an altitude at which there was no question of picking out a landing-place. We lost both wheels against one sand-dune, a wing against another, and crashed with a sudden jerk into a third.

"You hurt?" Riguelle called out.

"Not a bit," I said.

"That's what I call piloting a ship!" he boasted cheerfully.

I who was busy on all fours extricating myself from what had once been a ship, was in no mood to feed his pride.

"Guillaumet will be along in a minute to pick us up," he added.

Guillaumet was flying our convoy, and very shortly we saw him come down on a stretch of smooth sand a few hundred yards away. He asked if we were all right, was told no damage had been done, and then proposed briskly that we give him a hand with the sacks. The mail transferred out of the wrecked plane, they explained to me that in this soft sand it would not be possible to lift Guillaumet's plane clear if I was in it. They would hop to the next outpost, drop the mail there, and come back for me.

Now this was my first day in Africa. I was so ignorant that I could not tell a zone of danger from a zone of safety, I mean by that, a zone where the tribes had submitted peacefully to European rule from a zone where the tribes were still in rebellion. The region in which we had landed happened to be considered safe, but I did not know that.

"You've got a gun, of course," Riguelle said.

I had no gun and said so.

"My dear chap, you'll have to have a gun," he said, and very kindly he gave me his. "And you'll want these extra clips of cartridges," he went on. "Just bear in mind that you shoot at anything and everything you see."

They had started to walk across to the other plane when Guillaumet, as if driven by his conscience, came back and handed me his cartridge clips, too. And with this they took off.

I was alone. They knew, though I did not, that I could have sat on one of these dunes for half a year without running the least danger. What they were doing was to implant in the imagination of a recruit a proper feeling of solitude and danger and respect with regard to their desert. What I was really feeling, however, was an immense pride. Sitting on the dune, I laid out beside me my gun and my five cartridge clips. For the first time since I was born it seemed to me that my life was my own and that I was responsible for it. Bear in mind that only two nights before I had been dining in a restaurant in Toulouse.

I walked to the top of a sand-hill and looked round the horizon like a captain on his bridge. This sea of sand bowled me over. Unquestionably it was filled with mystery and with danger. The silence that reigned over it was not the silence of emptiness but of plotting, of imminent enterprise. I sat still and stared into space. The end of the day was near. Something half revealed yet wholly unknown had bewitched me. The love of the Sahara, like love itself, is born of a face perceived and never really seen. Ever after this first sight of your new love, an indefinable bond is established between you and the veneer of gold on the sand in the late sun.

Guillaumet's perfect landing broke the charm of my musings.

"Anything turn up?" he wanted to know.

I had seen my first gazelle. Silently it had come into view. I felt that the sands had shown me the gazelle in confidence, so I said nothing about it.

"You weren't frightened?"

I said no and thought, gazelles are not frightening.

The mails had been dropped at an outpost as isolated as an island in the Pacific. There, waiting for us, stood a colonial army sergeant. With his squad of fifteen black troops he stood guard on the threshold of the immense expanse. Every six months a caravan came up out of the desert and left him supplies.

Again and again he took our hands and looked into our eyes, ready to weep at the sight of us. "By God, I'm glad to see you! You don't know what it means to me to see you!" Only twice a year he saw a French face, and that was when, at the head of the camel corps, either the captain or the lieutenant came out of the inner desert.

We had to inspect his little fort—"built it with my own hands" —and swing his doors appreciatively—"as solid as they make 'em"—and drink a glass of wine with him.

"Another glass. Please! You don't know how glad I am to have some wine to offer you. Why, last time the captain came round I didn't have any for the captain. Think of that! I couldn't clink glasses with the captain and wish him luck! I was ashamed of myself. I asked to be relieved, I did!"

Clink glasses. Call out, "Here's luck!" to a man, running with sweat, who has just jumped down from the back of a camel. Wait six months for this great moment. Polish up your equipment. Scour the post from cellar to attic. Go up on the roof day after day and scan the horizon for that dust-cloud that serves as the envelope in which will be delivered to your door the Atar Camel Corps.

And after all this, to have no wine in the house! To be unable to clink glasses. To see oneself dishonored.

"I keep waiting for the captain to come back," the sergeant said.

"Where is he, sergeant?"

And the sergeant, waving his arm in an arc that took in the whole horizon, said: "Nobody knows. Captain is everywhere at once."

We spent the night on the roof of the outpost, talking about the stars. There was nothing else in sight. All the stars were present, all accounted for, the way you see them from a plane, but fixed.

When the night is very fine and you are at the stick of your ship, you half forget yourself and bit by bit the plane begins to tilt on the left. Pretty soon, while you still imagine yourself in plumb, you see the lights of a village under your right wing. There are no villages in the desert. A fishing-fleet in mid-ocean, then? There are no fishing-fleets in mid-Sahara. What—? Of course! You smile at the way your mind has wandered and you bring the ship back to plumb again. The village slips into place. You have hooked that particular constellation back in the panoply out of which it had fallen. Village? Yes, village of stars.

The sergeant had a word to say about them. "I know the stars," he said. "Steer by that star yonder and you make Tunis."

"Are you from Tunis?"

"No. My cousin, she is."

A long silence. But the sergeant could not keep anything back.

"I'm going to Tunis one of these days."

Not, I said to myself, by making a bee-line for that star and tramping across the desert; that is, not unless in the course of some raid a dried-up well should turn the sergeant over to the poetry of delirium. If that happened, star, cousin, and Tunis would melt into one, and the sergeant would certainly be off on that inspired tramp which the ignorant would think of as torture.

He went on. "I asked the captain for leave to go to Tunis, seeing my cousin is there and all. He said . . ."

"What did the captain say, sergeant?"

"Said: 'World's full of cousins.' Said: 'Dakar's nearer' and sent me there."

"Pretty girl, your cousin?"

"In Tunis? You bet! Blonde, she is."

"No, I mean at Dakar."

Sergeant, we could have hugged you for the wistful disappointed voice in which you answered, "She was a nigger."

II

Port Etienne is situated on the edge of one of the unsubdued regions of the Sahara. It is not a town. There is a stockade, a hangar, and a wooden quarters for the French crews. The desert all round is so unrelieved that despite its feeble military strength Port Etienne is practically invincible. To attack it means crossing such a belt of sand and flaming heat that the razzias (as the bands of armed marauders are called) must arrive exhausted and waterless. And yet, in the memory of man there has always been, somewhere in the North, a razzia marching on Port Etienne. Each time that the army captain who served as commandant of the fort came to drink a cup of tea with us, he would show us its route on the map the way a man might tell the legend of a beautiful princess.

But the razzia never arrived. Like a river, it was each time dried up by the sands, and we called it the phantom razzia. The cartridges and hand grenades that the government passed out to us nightly would sleep peacefully in their boxes at the foot of our beds. Our surest protection was our poverty, our single enemy silence. Night and day, Lucas, who was chief of the airport, would wind his gramophone; and Ravel's *Bolero,* flung up here so far out of the path of life, would speak to us in a half-lost language, provoking an aimless melancholy which curiously resembled thirst.

One evening we had dined at the fort and the commandant had shown off his garden to us. Someone had sent him from France, three thousand miles away, a few boxes of real soil, and out of this soil grew three green leaves which we caressed as if they had been jewels. The commandant would say of them, "This is my park." And when there arose one of those sand-storms that shriveled everything up, he would move the park down into the cellar.

Our quarters stood about a mile from the fort, and after dinner we walked home in the moonlight. Under the moon the sands were rosy. We were conscious of our destitution, but the sands were rosy. A sentry called out, and the pathos of our world was re-established. The whole of the Sahara lay in fear of our shadows and called for the password, for a razzia was on the march. All the voices of the desert resounded in that sentry's challenge. No longer was the desert an empty prison: a Moorish caravan had magnetized the night.

We might believe ourselves secure; and yet, illness, accident, razzia—how many dangers were afoot! Man inhabits the earth, a target for secret marksmen. The Senegalese sentry was there like a prophet of old to remind us of our destiny. We gave the password, *Français!* and passed before the black angel. Once in quarters, we breathed more freely. With what nobility that threat had endowed us! Oh, distant it still was, and so little urgent, deadened by so much sand; but yet the world was no longer the same. Once again this desert had become a sumptuous thing. A razzia that was somewhere on the march, yet never arrived, was the source of its glory.

It was now eleven at night. Lucas came back from the wireless and told me that the plane from Dakar would be in at midnight. All well on board. By ten minutes past midnight the mails would be transferred to my ship and I should take off for the North. I shaved carefully in a cracked mirror. From time to time, a Turkish towel hanging at my throat, I went to the door and looked at the naked sand. The night was fine but the wind was dropping. I went back again to the mirror. I was thoughtful.

A wind that has been running for months and then drops sometimes fouls the entire sky. I got into my harness, snapped my emergency lamps to my belt along with my altimeter and my pencils. I went over to Néri, who was to be my radio operator on this flight. He was shaving too. I said, "Everything all right?" For the moment everything was all right. But I heard something sizzling. It was a dragonfly knocking against the lamp. Why it was I cannot say, but I felt a twinge in my heart.

I went out of doors and looked round. The air was pure. A cliff on the edge of the airdrome stood in profile against the sky as if it were daylight. Over the desert reigned a vast silence as of a house in order. But here were a green butterfly and two dragonflies knocking against my lamp. And again I felt a dull ache which might as easily have been joy as fear but came up from the depths of me, so vague that it could scarcely be said to be there. Someone was calling to me from a great distance. Was it instinct?

Once again I went out. The wind had died down completely. The air was still cool. But I had received a warning. I guessed, I believed I could guess, what I was expecting. Was I right? Neither the sky nor the sand had made the least sign to me; but two dragonflies and a moth had spoken.

I climbed a dune and sat down face to the east. If I was right, the thing would not be long coming. What were they after here, those dragonflies, hundreds of miles from their oases inland? Wreckage thrown up on the strand bears witness to a storm at sea. Even so did these insects declare to me that a sand-storm was on the way, a storm out of the east that had blown them out of their oases.

Solemnly, for it was fraught with danger, the east wind rose. Already its foam had touched me. I was the extreme edge lapped by the wave. Fifty feet behind me no sail would have flapped. Its flame wrapped me round once, only once, in a caress that seemed dead. But I knew, in the seconds that followed, that the Sahara was catching its breath and would send forth a second sigh. And that before three minutes had passed the air-sock of our hangar would be whipped into action. And that before ten minutes had gone by the sand would fill the air. We should shortly be taking off in this conflagration, in this return of the flames from the desert.

But that was not what excited me. What filled me with a barbaric joy was that I had understood a murmured monosyllable of this secret language, had sniffed the air and known what was coming, like one of those primitive men to whom the future is revealed in such faint rustlings; it was that I had been able to read the anger of the desert in the beating wings of a dragonfly.

## III

But we were not always in the air, and our idle hours were spent taming the Moors. They would come out of their forbidden regions (those regions we crossed in our flights and where they would shoot at us the whole length of our crossing), would venture to the stockade in the hope of buying loaves of sugar, cotton cloth, tea, and then would sink back again into their mystery.

Whenever they turned up we would try to tame a few of them in order to establish little nuclei of friendship in the desert; thus if we were forced down among them there would be at any rate a few who might be persuaded to sell us into slavery rather than massacre us.

Now and then an influential chief came up, and him, with the approval of the Line, we would load into the plane and carry off to see something of the world. The aim was to soften their pride, for, repositories of the truth, defenders of Allah, the only God, it was more in contempt than in hatred that he and his kind murdered their prisoners.

When they met us in the region of Juby or Cisneros, they never troubled to shout abuse at us. They would merely turn away and spit; and this not by way of personal insult but out of sincere disgust at having crossed the path of a Christian. Their pride was born of the illusion of their power. Allah renders a believer invincible. Many a time a chief has said to me, pointing to his army of three hundred rifles, "Lucky it is for France that she lies more than a hundred days' march from here."

And so we would take them up for a little spin. Three of them even visited France in our planes. I happened to be present when they returned. I met them when they landed, went with them to their tents, and waited in infinite curiosity to hear their first words. They were of the same race as those who, having once been flown by me to the Senegal, had burst into tears at the sight of trees. What a revelation Europe must have been for them! And yet their first replies astonished me by their coolness.

"Paris? Very big."

Everything was "very big"—Paris, the Trocadéro, the automobiles.

What with everyone in Paris asking if the Louvre was not "very big" they had gradually learned that this was the answer that flattered us. And with a sort of vague contempt, as if pacifying a lot of children, they would grant that the Louvre was "very big."

These Moors took very little trouble to dissemble the freezing indifference they felt for the Eiffel Tower, the steamships, and the locomotives. They were ready to agree once and for always that we knew how to build things out of iron. We also knew how to fling a bridge from one continent to another. The plain fact was that they did not know enough to admire our technical progress. The wireless astonished them less than the telephone, since the mystery of the telephone resided in the very fact of the wire.

It took a little time for me to understand that my questions were on the wrong track. For what they thought admirable was not the locomotive, but the tree. When you think of it, a tree does possess a perfection that a locomotive cannot know. And then I remembered the Moors who had wept at the sight of trees.

Yes, France was in some sense admirable, but it was not because of those stupid things made of iron. They had seen pastures in France in which all the camels of Er-Reguibat could have grazed! There were forests in France! The French had cows, cows filled with milk! And of course my three Moors were amazed by the incredible customs of the people.

"In Paris," they said, "you walk through a crowd of a thousand people. You stare at them. And nobody carries a rifle!"

But there were better things in France than this inconceivable friendliness between men. There was the circus, for example.

"Frenchwomen," they said, "can jump standing from one galloping horse to another."

Thereupon they would stop and reflect.

"You take one Moor from each tribe," they went on. "You

take him to the circus. And nevermore will the tribes of Er-Reguibat make war on the French."

I remember my chiefs sitting among the crowding tribesmen in the opening of their tents, savoring the pleasure of reciting this new series of Arabian Nights, extolling the music halls in which naked women dance on carpets of flowers.

Here were men who had never seen a tree, a river, a rose; who knew only through the Koran of the existence of gardens where streams run, which is their name for Paradise. In their desert, Paradise and its beautiful captives could be won only by bitter death from an infidel's rifle-shot, after thirty years of a miserable existence. But God had tricked them, since from the Frenchmen to whom he grants these treasures he exacts payment neither by thirst nor by death. And it was upon this that the chiefs now mused. This was why, gazing out at the Sahara surrounding their tents, at that desert with its barren promise of such thin pleasures, they let themselves go in murmured confidences.

"You know . . . the God of the French . . . He is more generous to the French than the God of the Moors is to the Moors."

Memories that moved them too deeply rose to stop their speech. Some weeks earlier they had been taken up into the French Alps. Here in Africa they were still dreaming of what they saw. Their guide had led them to a tremendous waterfall, a sort of braided column roaring over the rocks. He had said to them:

"Taste this."

It was sweet water. Water! How many days were they wont to march in the desert to reach the nearest well; and when they had arrived, how long they had to dig before there bubbled a muddy liquid mixed with camel's urine! Water! At Cape Juby, at Cisneros, at Port Etienne, the Moorish children did not beg for coins. With empty tins in their hands they begged for water.

"Give me a little water, give!"

"If you are a good lad . . ."

Water! A thing worth its weight in gold! A thing the least drop of which drew from the sand the green sparkle of a blade of grass! When rain has fallen anywhere, a great exodus animates the Sahara. The tribes ride towards that grass that will have sprung up two hundred miles away. And this water, this miserly water of which not a drop had fallen at Port Etienne in ten years, roared in the Savoie with the power of a cataclysm as if, from some burst cistern, the reserves of the world were pouring forth.

"Come, let us leave," their guide had said.

But they would not stir.

"Leave us here a little longer."

They had stood in silence. Mute, solemn, they had stood gazing at the unfolding of a ceremonial mystery. That which came roaring out of the belly of the mountain was life itself, was the lifeblood of man. The flow of a single second would have resuscitated whole caravans that, mad with thirst, had pressed on into the eternity of salt lakes and mirages. Here God was manifesting Himself: It would not do to turn one's back on Him. God had opened the locks and was displaying His puissance. The three Moors had stood motionless.

"That is all there is to see," their guide had said. "Come."

"We must wait."

"Wait for what?"

"The end."

They were awaiting the moment when God would grow weary of His madness. They knew Him to be quick to repent, knew He was miserly.

"But that water has been running for a thousand years!"

And this was why, at Port Etienne, they did not too strongly stress the matter of the waterfall. There were certain miracles about which it was better to be silent. Better, indeed, not to think too much about them, for in that case one would cease to understand anything at all. Unless one was to doubt the existence of God. . . .

"You see . . . the God of the Frenchmen . . ."

But I knew them well, my barbarians. There they sat, perplexed in their faith, disconcerted, and henceforth quite ready to acknowledge French overlordship. They were dreaming of being victualed in barley by the French administration, and assured of their security by our Saharan regiments. There was no question but that they would, by their submission, be materially better off.

But all three were of the blood of el Mammun.

I had known el Mammun when he was our vassal. Loaded with official honors for services rendered, enriched by the French Government and respected by the tribes, he seemed to lack for nothing that belonged to the state of an Arab prince. And yet one night, without a sign of warning, he had massacred all the French officers in his train, had seized camels and rifles, and had fled to rejoin the refractory tribes in the interior.

Treason is the name given to these sudden uprisings, these flights at once heroic and despairing of a chieftain henceforth proscribed in the desert, this brief glory that will go out like a rocket against the low wall of European carbines. This sudden madness is properly a subject for amazement.

And yet the story of el Mammun was that of many other Arab chiefs. He grew old. Growing old, one begins to ponder. Pondering thus, el Mammun discovered one night that he had betrayed the God of Islam and had sullied his hand by sealing in the hand of the Christians a pact in which he had been stripped of everything.

Indeed what were barley and peace to him? A warrior disgraced and become a shepherd, he remembered a time when he had inhabited a Sahara where each fold in the sands was rich with hidden mysteries; where forward in the night the tip of the encampment was studded with sentries; where the news that spread concerning the movements of the enemy made all hearts beat faster round the night fires. He remembered a taste of the high seas which, once savored by man, is never forgotten. And because of his pact he was condemned to wander without glory through a region pacified and voided of all prestige. Then, truly and for the first time, the Sahara became a desert.

It is possible that he was fond of the officers he murdered. But love of Allah takes precedence.

"Good night, el Mammun."

"God guard thee!"

The officers rolled themselves up in their blankets and stretched out upon the sand as on a raft, face to the stars. High overhead all the heavens were wheeling slowly, a whole sky marking the hour. There was the moon, bending towards the sands, and the Frenchmen, lured by her tranquillity into oblivion, fell asleep. A few minutes more, and only the stars gleamed. And then, in order that the corrupted tribes be regenerated into their past splendor, in order that there begin again those flights without which the sands would have no radiance, it was enough that these Christians drowned in their slumber send forth a feeble wail. Still a few seconds more, and from the irreparable will come forth an empire.

And the handsome sleeping lieutenants were massacred.

## IV

Today at Cape Juby, Kemal and his brother Mouyan have invited me to their tent. I sit drinking tea while Mouyan stares at me in silence. Blue sandveil drawn across his mouth, he maintains an unsociable reserve. Kemal alone speaks to me and does the honors:

"My tent, my camels, my wives, my slaves are yours."

Mouyan, his eyes still fixed on me, bends towards his brother, pronounces a few words, and lapses into silence again.

"What does he say?" I ask.

"He says that Bonnafous has stolen a thousand camels from the tribes of Er-Reguibat."

I have never met this Captain Bonnafous, but I know that he is an officer of the camel corps garrisoned at Atar and I have gathered from the Moors that to them he is a legendary figure. They speak of him with anger, but as a sort of god. His presence lends price to the sand. Now once again, no one knows how, he has outflanked the southward marching razzias, taken them in the rear, driven off their camels by the hundred, and forced them to turn about and pursue him unless they are to lose those treasures which they had thought secure. And now, having saved Atar by this archangelic irruption and planted his camp upon a high limestone plateau, he stands there like a guerdon to be won, and such is his magnetism that the tribes are obliged to march towards his sword.

With a hard look at me, Mouyan speaks again.

"What now?" I ask.

"He says we are off tomorrow on a razzia against Bonnafous. Three hundred rifles."

I had guessed something of the sort. These camels led to the wells for three days past; these powwows; this fever running through the camp: it was as if men had been rigging an invisible ship. Already the air was filled with the wind that would take her out of port. Thanks to Bonnafous, each step to the South was to be a noble step rich in honor. It has become impossible to say whether love or hate plays the greater part in this setting forth of the warriors.

There is something magnificent in the possession of an enemy of Bonnafous' mettle. Where he turns up, the near-by tribes fold their tents, collect their camels and fly, trembling to think they might have found themselves face to face with him; while the more distant tribes are seized by a vertigo resembling love. They tear themselves from the peace of their tents, from the embraces of their women, from the happiness of slumber, for suddenly there is nothing in the world that can match in beauty, after two months of exhausting march, of burning thirst, of halts crouching under the sandstorm, the joy of falling unexpectedly at dawn upon the Atar camel corps and there, God willing, killing Captain Bonnafous.

"Bonnafous is very clever," Kemal avows.

Now I know their secret. Even as men who desire a woman dream of her indifferent footfall, toss and turn in the night, scorched and wounded by the indifference of that stroll she takes through their dream, so the distant progress of Bonnafous torments these warriors.

This Christian in Moorish dress at the head of his two hundred maurauding cameleers, Moors themselves, outflanking the razzias hurled against him, has marched boldly into the country of the refractory tents where the least of his own men, freed from the constraint of the garrison, might with impunity shake off his servitude and sacrifice the captain to his God on the stony table-lands. He has gone into a world where only his prestige restrains his men, where his weakness itself is the cause of their dread. And tonight, through their raucous slumber he strolls to and fro with heedless step, and his footfall resounds in the innermost heart of the desert.

Mouyan ponders, still motionless against the back wall of the tent, like a block of blue granite cut in low relief. Only his eyes gleam, and his silver knife has ceased to be a plaything. I have the feeling that since becoming part of a razzia he has entered a different world. To him the dunes are alive. The wind is charged with odors. He senses as never before his own nobility and crushes me beneath his contempt; for he is to ride against Bonnafous, he is to move at dawn impelled by a hatred that bears all the signs of love.

Once again he leans towards his brother, whispers, and stares at me.

"What is he saying?" I ask once again.

"That he will shoot you if he meets you outside the fort."

"Why?"

"He says you have airplanes and the wireless; you have Bonnafous; but you have not the Truth."

Motionless in the sculptured folds of his blue cloak, Mouyan has judged me.

"He says you eat greens like the goat and pork like the pigs. Your wives are shameless and show their faces—he has seen them. He says you never pray. He says, what good are your airplanes and wireless and Bonnafous, if you do not possess the Truth?"

And I am forced to admire this Moor who is not about to defend his freedom, for in the desert a man is always free; who is not about to defend his visible treasures, for the desert is bare; but who is about to defend a secret kingdom.

In the silence of the sand-waves Bonnafous leads his troop like a corsair of old; by the grace of Bonnafous the oasis of Cape Juby has ceased to be a haunt of idle shepherds and has become something as signal, as portentous, as admirable as a ship on the high seas. Bonnafous is a storm beating against the ship's side, and because of him the tent cloths are closed at night. How poignant is the southern silence! It is Bonnafous' silence. Mouyan, that old hunter, listens to his footfall in the wind.

When Bonnafous returns to France his enemies, far from rejoicing, will bewail his absence, as if his departure had deprived the desert of one of its magnetic poles and their existence of a part of its prestige. They will say to me:

"Why does Bonnafous leave us?"

"I do not know."

For years he had accepted their rules as his rules. He had staked his life against theirs. He had slept with his head pillowed on their rocks. Like them he had known Biblical nights of stars and wind in the course of the ceaseless pursuit. And of a sudden he proves to them, by the fact of leaving the desert, that he has not been gambling for a stake he deemed essential. Unconcernedly, he throws in his hand and rises from the table. And those Moors he leaves at their gambling lose confidence in the significance of a game which does not involve this man to the last drop of his blood. Still, they try to believe in him:

"Your Bonnafous will come back."

"I do not know."

He will come back, they tell themselves. The games of Europe will never satisfy him—garrison bridge, promotion, women, and the rest. Haunted by his lost honor he will come back to this land where each step makes the heart beat faster like a step towards love or towards death. He had imagined that the Sahara was a mere adventure and that what was essential in life lay in Europe; but he will discover with disgust that it was here in the desert he possessed his veritable treasures—this prestige of the sand, the night, the silence, this homeland of wind and stars.

And if Bonnafous should come back one day, the news will spread in a single night throughout the country of the refractory tribes. The Moors will know that somewhere in the Sahara, at the head of his two hundred marauders, Bonnafous is again on the march. They will lead their dromedaries in silence to the wells. They will prepare their provisions of barley. They will clean and oil their breech-loaders, impelled by a hatred that partakes of love.

## V

"Hide me in the Marrakech plane!"

Night after night, at Cape Juby, this slave would make his prayer to me. After which, satisfied that he had done what he could for his salvation, he would sit down upon crossed legs and brew my tea. Having put himself in the hands of the only doctor (as he believed) who could cure him, having prayed to the only god who might save him, he was at peace for another twenty-four hours.

Squatting over his kettle, he would summon up the simple vision of his past—the black earth of Marrakech, the pink houses, the rudimentary possessions of which he had been despoiled. He bore me no ill-will for my silence, nor for my delay in restoring him to life. I was not a man like himself but a power to be invoked, something like a favorable wind which one of these days might smile upon his destiny.

I, for my part, did not labor under these delusions concerning my power. What was I but a simple pilot, serving my few months as chief of the airport at Cape Juby and living in a wooden hut built over against the Spanish fort, where my worldly goods consisted of a basin, a jug of brackish water, and a cot too short for me?

"We shall see, Bark."

All slaves are called Bark, so Bark was his name. But despite four years of captivity he could not resign himself to it and remembered constantly that he had been a king.

"What did you do at Marrakech, Bark?"

At Marrakech, where his wife and three children were doubtless still living, he had plied a wonderful trade.

"I was a drover, and my name was Mohammed!"

The very magistrates themselves would send for him.

"Mohammed, I have some steers to sell. Go up into the mountains and bring them down."

Or:

"I have a thousand sheep in the plain. Lead them up into the higher pastures."

And Bark, armed with an olive-wood sceptre, governed their exodus. He and no other held sway over the nation of ewes, restrained the liveliest because of the lambkins about to be born, stirred up the laggards, strode forward in a universe of confidence and obedience. Nobody but him could say where lay the promised land towards which he led his flock. He alone could read his way in the stars, for the science he possessed was not shared by the sheep. Only he, in his wisdom, decided when they should take their rest, when they should drink at the springs. And at night while they slept, Bark, physician and prophet and king, standing in wool to the knees and swollen with tenderness for so much feeble ignorance, would pray for his people.

One day he was stopped by some Arabs.

"Come with us to fetch cattle up from the South," they said.

They had walked him a long time, and when, after three days, they found themselves deep in the mountains, on the borders of rebellion, the Arabs had quietly placed a hand on his shoulder, christened him Bark, and sold him into slavery.

He was not the only slave I knew. I used to go daily to the tents to take tea. Stretched out with naked feet on the thick woolen carpet which is the nomad's luxury and upon which for a time each day he builds his house, I would taste the happiness of the journeying hours. In the desert, as on shipboard, one is sensible of the passage of time. In that parching heat a man feels that the day is a voyage towards the goal of evening, towards the promise of a cool breeze that will bathe the limbs and wash away the sweat. Under the heat of the day beasts and men plod towards the sweet well of night as confidently as towards death. Thus, idleness here is never vain; and each day seems as comforting as the roads that lead to the sea.

I knew the slaves well. They would come in as soon as the chief had taken out the little stove, the kettle, and the glasses from his treasure chest—that chest heavy with absurd objects, with locks lacking keys, vases for non-existent flowers, threepenny mirrors, old weapons, things so disparate that they might have been salvaged from a ship cast up here in the desert.

Then the mute slave would cram the stove with twigs, blow on the embers, fill the kettle with water, and in this service that a child could perform, set into motion a play of muscles able to uproot a tree.

I would wonder what he was thinking of, and would sense that he was at peace with himself. There was no doubt that he was hypnotized by the motions he went through—brewing tea, tending the camels, eating. Under the blistering day he walked towards the night; and under the ice of the naked stars he longed for the return of day. Happy are the lands of the North whose seasons are poets, the summer composing a legend of snow, the winter a tale of sun. Sad the tropics, where in the sweating-room nothing changes very much. But happy also the Sahara where day and night swing man so evenly from one hope to the other.

Tea served, the black will squat outside the tent, relishing the evening wind. In this sluggish captive hulk, memories have ceased to swarm. Even the moment when he was carried off is faint in his mind—the blows, the shouts, the arms of men that brought him down into his present night. And since that hour he has sunk deeper and deeper into a queer slumber, divested like a blind man of his Senegalese rivers or his white Moroccan towns, like a deaf man of the sound of familiar voices.

This black is not unhappy; he is crippled. Dropped down one day into the cycle of desert life, bound to the nomadic migrations, chained for life to the orbits they describe in the sand, how could he retain any memory of a past, a home, a wife and children, all of them for him as dead as the dead?

Men who have lived for years with a great love, and have lived on in noble solitude when it was taken from them, are likely now and then to be worn out by their exaltation. Such men return humbly to a humdrum life, ready to accept contentment in a more commonplace love. They find it sweet to abdicate, to resign themselves to a kind of servility and to enter into the peace of things. This black is proud of his master's embers.

Like a ship moving into port, we of the desert come up into the night. In this hour, because it is the hour when all the weariness of day is remitted and its heats have ceased, when master and slave enter side by side into the cool of evening, the master is kind to the slave.

"Here, take this," the chief says to the captive.

He allows him a glass of tea. And the captive, overcome with gratitude for a glass of tea, would kiss his master's knees. This man before me is not weighed down with chains. How little need he has of them! How faithful he is! How submissively he forswears the deposed king within him! Truly, the man is a mere contented slave.

And yet the day will come when he will be set free. When he has grown too old to be worth his food or his cloak, he will be inconceivably free. For three days he will offer himself in vain from tent to tent, growing each day weaker; until towards the end of the third day, still uncomplaining, he will lie down on the sand.

I have seen them die naked like this at Cape Juby. The Moors jostle their long death-struggle, though without ill intent; and the children play in the vicinity of the dark wreck, running with each dawn to see if it is still stirring, yet without mocking the old servitor. It is all in the nature of things. It is as if they had said to him: "You have done a good day's work and have the right to sleep. Go to bed."

And the old slave, still outstretched, suffers hunger which is but vertigo, and not injustice which alone is torment. Bit by bit he becomes one with the earth, is shriveled up by the sun and received by the earth. Thirty years of toil, and then this right to slumber and to the earth.

The first one I saw did not moan; but then he had no one to moan against. I felt in him an obscure acquiescence, as of a mountaineer lost and at the end of his strength who sinks to earth and wraps himself up in dreams and snow. What was painful to me was not his suffering (for I did not believe he was suffering); it was that for the first time it came on me that when a man dies, an unknown world passes away.

I could not tell what visions were vanishing in the dying slave, what Senegalese plantations or white Moroccan towns. It was impossible for me to know whether, in this black heap, there was being extinguished merely a world of petty cares in the breast of a slave—the tea to be brewed, the camels watered; or whether, revived by a surge of memories, a man lay dying in the glory of humanity. The hard bone of his skull was in a sense an old treasure chest; and I could not know what colored stuffs, what images of festivities, what vestiges, obsolete and vain in this desert, had here escaped the shipwreck.

The chest was there, locked and heavy. I could not know what bit of the world was crumbling in this man during the gigantic slumber of his ultimate days, was disintegrating in this consciousness and this flesh which little by little was reverting to night and to root.

"I was a drover, and my name was Mohammed!"

Before I met Bark I had never met a slave who offered the least resistance. That the Moors had violated his freedom, had in a single day stripped him as naked as a new-born infant, was not the point. God sometimes sends cyclones which in a single hour wipe out a man's harvests. But deeper than his belongings, these Moors had threatened him in his very essence.

Many another captive would have resigned himself to the death in him of the poor herdsman who toiled the year round for a crust of bread. Not so Bark. He refused to settle into a life of servitude, to surrender to the weariness of waiting and resign himself to a passive contentment. He rejected the slave-joys that are contingent upon the kindness of the slave-owner. Within his breast Mohammed absent held fast to the house Mohammed had lived in. That house was sad for being empty, but none other should live in it. Bark was like one of those white-haired caretakers who die of their fidelity in the weeds of the paths and the tedium of silence.

He never said, "I am Mohammed ben Lhaoussin"; he said, "My name was Mohammed," dreaming of the day when that obliterated figure would again live within him in all its glory and by the power of its resuscitation would drive out the ghost of the slave.

There were times when, in the silence of the night, all his memories swept over him with the poignancy of a song of childhood. Our Arab interpreter said to me, "In the middle of the night he woke up and talked about Marrakech; and he wept." No man in solitude can escape these recurrences. The old Mohammed awoke in him without warning, stretched himself in his limbs, sought his wife against his flank in this desert where no woman had ever approached Bark, and listened to the water purling in the fountains here where no fountain ran.

And Bark, his eyes shut, sitting every night under the same star, in a place where men live in houses of hair and follow the wind, told himself that he was living in his white house in Marrakech. His body charged with tenderness and mysteriously magnetized, as if the pole of these emotions were very near at hand, Bark would come to see me. He was trying to let me know that he was ready, that his over-full heart was quivering on the brim and needed only to find itself back in Marrakech to be poured out. And all that was wanted was a sign from me. Bark would smile, would whisper to me how it could be done—for of course I should not have thought of this dodge:

"The mails leave tomorrow. You stow me away in the Marrakech plane."

"Poor old Bark!"

We were stationed among the unsubdued tribes, and how could we help him away? God knows what massacre the Moors would have done among us that very day to avenge the insult of this theft. I had, indeed, tried to buy him, with the help of the mechanics at the port—Laubergue, Marchal, and Abgrall. But it was not every day that the Moors met Europeans in quest of a slave, and they took advantage of the occasion.

"Twenty thousand francs."

"Don't make me laugh!"

"But look at those strong arms. . . ."

Months passed before the Moors came down to a reasonable figure and I, with the help of friends at home to whom I had written, found myself in a position to buy old Bark. There was a week of bargaining which we spent, fifteen Moors and I, sitting in a circle in the sand. A friend of Bark's master who was also my friend, Zin Ould Rhattari, a bandit, was privately on my side.

"Sell him," he would argue in accordance with my coaching. "You will lose him one of these days, you know. Bark is a sick man. He is diseased. You can't see yet, but he is sick inside. One of these days he will swell right up. Sell him as soon as you can to the Frenchman."

I had promised fifty Spanish pesetas to another bandit, Raggi, and Raggi would say:

"With the money you get for Bark you will be able to buy camels and rifles and cartridges. Then you can go off on a razzia against these French. Go down to Atar and bring back three or four young Senegalese. Get rid of the old carcass."

And so Bark was sold to me. I locked him up for six days in our hut, for if he had wandered out before the arrival of a plane the Moors would surely have kidnapped him. Meanwhile, although I would not allow him out, I set him free with a flourish of ceremony in the presence of three Moorish witnesses. One was a local marabout, another was Ibrahim, the mayor of Cape Juby, and the third was his former owner. These three pirates, who would gladly have cut off Bark's head within fifty feet of the fort for the sole pleasure of doing me in the eye, embraced him warmly and signed the official act of manumission. That done, they said to him:

"You are now our son."

He was my son, too, by law. Dutifully, Bark embraced all his fathers.

He lived on in our hut in comfortable captivity until we could ship him home. Over and over again, twenty times a day, he would ask to have the simple journey described. We were flying him to Agadir. There he would be given an omnibus ticket to Marrakech. He was to be sure not to miss the bus. That was all there was to it. But Bark played at being free the way a child plays at being an explorer, going over and over this journey back to life—the bus, the crowds, the towns he would pass through.

One day Laubergue came to talk to me about Bark. He said that Marchal and Abgrall and he rather felt it would be a shame if Bark was flung into the world without a copper. They had made up a purse of a thousand francs: didn't I think that would see Bark through till he found work? I thought of all the old ladies who run charities and insist upon gratitude in exchange for every twenty francs they part with. These airplane mechanics were parting with a thousand francs, had no thought of charity, and were even less concerned about gratitude.

Nor were they acting out of pity, like those old ladies who want to believe they are spreading happiness. They were contributing simply to restore to a man his lost dignity as a human being. They knew quite as well as anybody else that once the initial intoxication of his homecoming was past, the first faithful friend to step up and take Bark's hand would be Poverty; and that before three months had gone by he would be tearing up sleepers somewhere on the railway line for a living. He was sure to be less well off there than here in the desert. But in their view he had the right to live his life among his own people.

"Good-by, old Bark. Be a man!"

The plane quivered, ready to take off. Bark took his last look at the immense desolation of Cape Juby. Round the plane two hundred Moors were finding out what a slave looked like when he stood on the threshold of life. They would make no bones about snatching him back again if a little later the ship happened to be forced down.

We stood about our fifty-year-old, new-born babe, worried a little at having launched him forth on the stream of life.

"Good-by, Bark!"

"No!"

"What do you mean?"

"No. I am Mohammed ben Lhaoussin."

The last news we had of him was brought back to us by Abdullah who at our request had looked after Bark at Agadir. The plane reached Agadir in the morning, but the bus did not leave until evening. This was how Bark spent his day.

He began by wandering through the town and remaining silent so long that his restlessness upset Abdullah.

"Anything the matter?"

"No."

This freedom had come too suddenly: Bark was finding it hard to orient himself. There was a vague happiness in him, but with this exception there was scarcely any difference between the Bark of yesterday and the Bark of today. Yet he had as much right to the sun, henceforth, as other men; as much right as they to sit in the shade of an Arab café.

He sat down and ordered tea for Abdullah and himself. This was his first lordly gesture, a manifestation of a power that ought to have transfigured him in other men's eyes. But the waiter poured his tea quite without surprise, quite unaware that in this gesture he was doing homage to a free man.

"Let us go somewhere else," Bark had said; and they had gone off to the Kasbah, the licensed quarter of the town. The little Berber prostitutes came up and greeted them, so kind and tame that here Bark felt he might be coming alive.

These girls were welcoming a man back to life, but they knew nothing of this. They took him by the hand, offered him tea, then love, very nicely; but exactly as they would have offered it to any man. Bark, preoccupied with his message, tried to tell them the story of his resurrection. They smiled most sympathetically. They were glad for him, since he was glad. And to make the wonder more wonderful he added, "I am Mohammed ben Lhaoussin."

But that was no surprise to them. All men have names, and so many return from afar! They could guess, nevertheless, that this man had suffered, and they strove to be as gentle as possible with the poor black devil. He appreciated their gentleness, this first gift that life was making him; but his restlessness was yet not stilled. He had not yet rediscovered his empire.

Back to town went Bark and Abdullah. He idled in front of the Jewish shops, stared at the sea, repeated to himself that he could walk as he pleased in any direction, that he was free. But this freedom had in it a taste of bitterness: what he learned from it with most intensity was that he had no ties with the world.

At that moment a child had come up. Bark stroked the soft cheek. The child smiled. This was not one of the master's children that one had to flatter. It was a sickly child whose cheek Bark was stroking. And the child was smiling at him. The child awoke something in Bark, and Bark felt himself more important on earth because of the sickly child whose smile was his due. He began to sense confusedly that something was stirring within him, was striding forward with swift steps.

"What are you looking for?" Abdullah had asked him.

"Nothing," was again Bark's answer.

But when, rounding a corner, he came upon a group of children at play, he stopped. This was it. He stared at them in silence. Then he went off to the Jewish shops and came back laden with treasure. Abdullah was nettled:

"Fool! Throwing away your money!"

Bark gave no heed. Solemnly he beckoned to each child in turn, and the little hands rose towards the toys and the bangles and the gold-sewn slippers. Each child, as soon as he had a firm grip on his treasure, fled like a wild thing, and Bark went back to the Jewish shops.

Other children in Agadir, hearing the news, ran after him; and these too were shod by Bark in golden slippers. The tale spread to the outskirts of Agadir, whence still other children scurried into town and clustered round the black god, clinging to his threadbare cloak and clamoring for their due. Bark, that victim of a sombre joy, spent on them his last copper.

Abdullah was sure that he had gone mad, "mad with joy," he said afterward. But I incline to believe that Bark was not sharing with others an overflow of happiness. He was free, and therefore he possessed the essential of wealth—the right to the love of Berber girls, to go north or south as he pleased, to earn his bread by his toil. What good was this money when the thing for which he was famished was to be a man in the family of men, bound by ties to other men?

The town prostitutes had been kind to old Bark, but he had been able to get away from them as easily as he had come to them: they had no need of him. The waiter in the café, the passers-by in the streets, the shopkeepers, had respected the free man he was, sharing their sun with him on terms of equality; but none of them had indicated that he needed Bark.

He was free, but too infinitely free; not striding upon the earth but floating above it. He felt the lack in him of that weight of human relations that trammels a man's progress; tears, farewells, reproaches, joys—all those things that a man caresses or rips apart each time he sketches a gesture; those thousand ties that bind him to others and lend density to his being. But already Bark was in ballast of a thousand hopes.

And so the reign of Bark began in the glory of the sun setting over Agadir, in that evening coolness that so long had been for him the single sweetness, the unique stall in which he could take his rest. And as the hour of leaving approached, Bark went forward lapped in this tide of children as once in his sea of ewes, ploughing his first furrow in the world. He would go back next day to the poverty of his family, to responsibility for more lives than perhaps his old arms would be able to sustain; but already, among these children, he felt the pull of his true weight. Like an archangel too airy to live the life of man, but who had cheated, had sewn lead into his girdle, Bark dragged himself forward, pulling against the pull of a thousand children who had such great need of golden slippers.

Such is the desert. A Koran which is but a handbook of the rules of the game transforms its sands into an empire. Deep in the seemingly empty Sahara a secret drama is being played that stirs the passions of men. The true life of the desert is not made up of the marches of tribes in search of pasture, but of the game that goes endlessly on. What a difference in substance between the sands of submission and the sands of unruliness! The dunes, the salines, change their nature according as the code changes by which they are governed.

And is not all the world like this? Gazing at this transfigured desert I remember the games of my childhood—the dark and golden park we peopled with gods; the limitless kingdom we made of this square mile never thoroughly explored, never thoroughly charted. We created a secret civilization where footfalls had a meaning and things a savor known in no other world.

And when we grow to be men and live under other laws, what remains of that park filled with the shadows of childhood, magical, freezing, burning? What do we learn when we return to it and stroll with a sort of despair along the outside of its little wall of gray stone, marveling that within a space so small we should have founded a kingdom that had seemed to us infinite—what do we learn except that in this infinity we shall never again set foot, and that it is into the game and not the park that we have lost the power to enter?

# VIII

## *Prisoner of the Sand*

After three years of life in the desert, I was transferred out. The fortunes of the air service sent me wandering here and there until one day I decided to attempt a long-distance flight from Paris to

Saïgon. When, on December 29, 1935, I took off, I had no notion that the sands were preparing for me their ultimate and culminating ordeal.

This is the story of the Paris-Saïgon flight.

I paid my final visit to the weather bureau, where I found Monsieur Viaud stooped over his maps like a medieval alchemist over an alembic. Lucas had come with me, and we stared together at the curving lines marking the new-sprung winds. With their tiny flying arrows, they put me in mind of curving tendrils studded with thorns. All the atmospheric depressions of the world were charted on this enormous map, ochre-colored, like the earth of Asia.

"Here is a storm that we'll not hear from before Monday," Monsieur Viaud pointed out.

Over Russia and the Scandinavian peninsula the swirling lines took the form of a coiled demon. Out in Iraq, in the neighborhood of Basra, an imp was whirling.

"That fellow worries me a little," said Monsieur Viaud.

"Sand-storm, is it?"

I was not being idly curious. Day would not yet be breaking when I reached Basra and I was fearful of flying at night in one of those desert storms that turn the sky into a yellow furnace and wipe out hills, towns, and river-banks, drowning earth and sky in one great conflagration. It would be bad enough to fly in daylight through a chaos in which the very elements themselves were indistinguishable.

"Sand-storm? No, not exactly."

"So much the better," I said to myself, and I looked round the room. I liked this laboratory atmosphere. Viaud, I felt, was a man escaped from the world. When he came in here and hung up his hat and coat on the peg, he hung up with them all the confusion in which the rest of mankind lived. Family cares, thoughts of income, concerns of the heart—all that vanished on the threshold of this room as at the door of a hermit's cell, or an astronomer's tower, or a radio operator's shack. Here was one of those men who are able to lock themselves up in the secrecy of their retreat and hold discourse with the universe.

Gently, for he was reflecting, Monsieur Viaud rubbed the palms of his hands together.

"No, not a sand-storm. See here."

His finger traveled over the map and pointed out why.

At four in the morning Lucas shook me into consciousness.

"Wake up!"

And before I could so much as rub my eyes he was saying, "Look here, at this report. Look at the moon. You won't see much of her tonight. She's new, not very bright, and she'll set at ten o'clock. And here's something else for you: sunrise in Greenwich Meridian Time and in local time as well. And here: here are your maps, with your course all marked out. And here—"

"—is your bag packed for Saïgon," my wife broke in.

A razor and a change of shirt. He who would travel happily must travel light.

We got into a car and motored out to Le Bourget while Fate spying in ambush put the finishing touches to her plans. Those favorable winds that were to wheel in the heavens, that moon that was to sink at ten o'clock, were so many strategic positions at which Fate was assembling her forces.

It was cold at the airport, and dark. The *Simoon* was wheeled out of her hangar. I walked round my ship, stroking her wings with the back of my hand in a caress that I believe was love. Eight thousand miles I had flown in her, and her engines had not skipped a beat; not a bolt in her had loosened. This was the marvel that was to save our lives the next night by refusing to be ground to powder on meeting the upsurging earth.

Friends had turned up. Every long flight starts in the same atmosphere, and nobody who has experienced it once would ever have it otherwise: the wind, the drizzle at daybreak, the engines purring quietly as they are warmed up; this instrument of conquest gleaming in her fresh coat of "dope"—all of it goes straight to the heart.

Already one has a foretaste of the treasures about to be garnered on the way—the green and brown and yellow lands promised by the maps; the rosary of resounding names that make up the pilot's beads; the hours to be picked up one by one on the eastward flight into the sun.

There is a particular flavor about the tiny cabin in which, still only half awake, you stow away your thermos flasks and odd parts and over-night bag; in the fuel tanks heavy with power; and best of all, forward, in the magical instruments set like jewels in their panel and glimmering like a constellation in the dark of night. The mineral glow of the artificial horizon, these stethoscopes designed to take the heart-beat of the heavens, are things a pilot loves. The cabin of a plane is a world unto itself, and to the pilot it is home.

I took off, and though the load of fuel was heavy, I got easily away. I avoided Paris with a jerk and up the Seine, at Melun, I found myself flying very low between showers of rain. I was heading for the valley of the Loire. Nevers lay below me, and then Lyon. Over the Rhône I was shaken up a bit. Mt. Ventoux was capped in snow. There lies Marignane and here comes Marseille.

The towns slipped past as in a dream. I was going so far—or thought I was going so far—that these wretched little distances were covered before I was aware of it. The minutes were flying. So much the better. There are times when, after a quarter-hour of flight, you look at your watch and find that five minutes have gone by; other days when the hands turn a quarter of an hour in the wink of an eye. This was a day when time was flying. A good omen. I started out to sea.

Very odd, that little stream of vapor rising from the fuel gauge on my port wing! It might almost be a plume of smoke.

"Prévot!"

My mechanic leaned towards me.

"Look! Isn't that gas? Seems to me it's leaking pretty fast."

He had a look and shook his head.

"Better check our consumption," I said.

I wasn't turning back yet. My course was still set for Tunis. I looked round and could see Prévot at the gauge on the second fuel tank aft. He came forward and said:

"You've used up about fifty gallons."

Nearly twenty had leaked away in the wind! That was serious. I put back to Marignane where I drank a cup of coffee while the time lost hurt like an open wound. Flyers in the Air France service wanted to know whether I was bound for Saïgon or Madagascar and wished me luck. The tank was patched up and refilled, and I took off once more with a full load, again without mishap despite a bit of rough going over the soggy field.

As soon as I reached the sea I ran into low-hanging clouds that forced me down to sixty feet. The driving rain spattered against the windshield and the sea was churning and foaming. I strained to see ahead and keep from hooking the mast of some ship, while Prévot lit cigarettes for me.

"Coffee!"

He vanished into the stern of the cockpit and came back with the thermos flask. I drank. From time to time I flicked the throttle to keep the engines at exactly 2100 revolutions and ran my eye over the dials like a captain inspecting his troops. My company stood trim and erect: every needle was where it should be.

I glanced down at the sea and saw it bubbling under the steaming rain like a boiling cauldron. In a hydroplane this bumpy sea would have bothered me; but in this ship of mine, which could not possibly be set down here, I felt differently. It was silly, of course, but the thought gave me a sense of security. The sea was part of a world that I had nothing to do with. Engine trouble here was out of the question: there was not the least danger of such a thing. Why, I was not rigged for the sea!

After an hour and a half of this, the rain died down, and though the clouds still hung low a genial sun began to break through. I was immensely cheered by this promise of good weather. Overhead I could feel a thin layer of cotton-wool and I swerved aside to avoid a downpour. I was past the point where I had to cut through the heart of squalls. Was not that the first rift in the cloud-bank, there ahead of me?

I sensed it before I saw it, for straight ahead on the sea lay a long meadow-colored swath, a sort of oasis of deep and luminous green reminding me of those barley fields in southern Morocco that would make me catch my breath each time I sighted them on coming up from Senegal across two thousand miles of sand. Here as at such times in Morocco I felt we had reached a place a man could live in, and it bucked me up. I flung a glance backward at Prévot and called out:

"We're over the worst of it. This is fine."

"Yes," he said, "fine."

This meant that I would not need to do any stunt flying when Sardinia hove unexpectedly into view. The island would not loom up suddenly like a mass of wreckage a hundred feet ahead of me: I should be able to see it rising on the horizon in the distant play of a thousand sparkling points of light.

I moved into this region bathed by the sun. No doubt about it, I was loafing along. Loafing at the rate of one hundred and seventy miles an hour, but loafing nevertheless. I smoked a few leisurely cigarettes. I lingered over my coffee. I kept a cautious fatherly eye on my brood of instruments. These clouds, this sun, this play of light, lent to my flight the relaxation of a Sunday afternoon stroll. The sea was as variegated as a country landscape broken into fields of green and violet and blue. Off in the distance, just where a squall was blowing, I could see the fermenting spray. Once again I recognized that the sea was of all things in the world the least monotonous, was formed of an ever-changing substance. A gust of wind mantles it with light or strips it bare. I turned back to Prévot.

"Look!" I said.

There in the distance lay the shores of Sardinia that we were about to skirt to the southward.

Prévot came forward and sat down beside me. He squinted with wrinkled forehead at the mountains struggling out of their shroud of mist. The clouds had been blown away and the island was coming into view in great slabs of field and woodland. I climbed to forty-five hundred feet and drifted along the coast of this island dotted with villages. After the flower-strewn but uninhabitable sea, this was a place where I could take things easily. For a little time I clung to our great-hearted mother earth. Then, Sardinia behind me, I headed for Tunis.

I picked up the African continent at Bizerta and there I began to drop earthward. I was at home. Here was a place where I could dispense with altitude which, as every pilot knows, is our particular store of wealth. Not that we squander it when it is no longer needed: we swap it for another kind of treasure. When a flyer is within a quarter of an hour of port, he sets his controls for the down swing, throttling his motor a little—just enough to keep it from racing while the needle on his speedometer swings round from one hundred and seventy to two hundred miles an hour.

At that rate of speed the impalpable eddies of evening air drum softly on the wings and the plane seems to be drilling its way into a quivering crystal so delicate that the wake of a passing swallow would jar it to bits. I was already skirting the undulations of the hills and had given away almost the whole of my few hundred feet of altitude when I reached the airdrome, and there, shaving the roofs of the hangars, I set down my ship on the ground.

While the tanks were being re-filled I signed some papers and shook hands with a few friends. And just as I was coming out of the administration building I heard a horrible grunt, one of those muffled impacts that tell their fatal story in a single sound; one of those echoless thuds complete in themselves, without appeal, in which fatality delivers its message. Instantly there came into my mind the memory of an identical sound—an explosion in a garage. Two men had died of that hoarse bark.

I looked now across to the road that ran alongside the airdrome: there in a puff of dust two high-powered cars had crashed head-on and stood frozen into motionlessness as if imprisoned in ice. Men were running towards the cars while others ran from them to the field office.

"Get a doctor. . . . Skull crushed. . . ."

My heart sank. In the peace of the evening light Fate had taken a trick. A beauty, a mind, a life—something had been destroyed. It was as sudden as a raid in the desert. Marauding tribesmen creep up on silent feet in the night. The camp resounds briefly with the clashing tumult of a razzia. A moment later everything has sunk back into the golden silence. The same peace, the same stillness, followed this crash.

Near by, someone spoke of a fractured skull. I had no mind to be told about that crushed and bloody cranium. Turning my back to the road, I went across to my ship, in my heart a foreboding of danger. I was to recognize that sound when I heard it again very soon. When the *Simoon* scraped the black plateau at a speed of one hundred and seventy miles an hour I should recognize that hoarse grunt, that same snarl of destiny keeping its appointment with us.

Off to Benghazi! We still have two hours of daylight. Before we crossed into Tripolitana I took off my glare glasses. The sands were golden under the slanting rays of the sun. How empty of life is this planet of ours! Once again it struck me that its rivers, its woods, its human habitations were the product of chance, of fortuitous conjunctions of circumstance. What a deal of the earth's surface is given over to rock and sand!

But all this was not my affair. My world was the world of flight. Already I could feel the oncoming night within which I should be enclosed as in the precincts of a temple—enclosed in the temple of night for the accomplishment of secret rites and absorption in inviolable contemplation.

Already this profane world was beginning to fade out: soon it would vanish altogether. This landscape was still laved in golden sunlight, but already something was evaporating out of it. I know nothing, nothing in the world, equal to the wonder of nightfall in the air.

Those who have been enthralled by the witchery of flying will know what I mean—and I do not speak of the men who, among other sports, enjoy taking a turn in a plane. I speak of those who fly professionally and have sacrificed much to their craft. Mermoz said once, "It's worth it, it's worth the final smash-up."

No question about it; but the reason is hard to formulate. A novice taking orders could appreciate this ascension towards the essence of things, since his profession too is one of renunciation: he renounces the world; he renounces riches; he renounces the love of woman. And by renunciation he discovers his hidden god.

I, too, in this flight, am renouncing things. I am giving up the broad golden surfaces that would befriend me if my engines were to fail. I am giving up the landmarks by which I might be taking my bearings. I am giving up the profiles of mountains against the sky that would warn me of pitfalls. I am plunging into the night. I am navigating. I have on my side only the stars.

The diurnal death of the world is a slow death. It is only little by little that the divine beacon of daylight recedes from me. Earth and sky begin to merge into each other. The earth rises and seems to spread like a mist. The first stars tremble as if shimmering in green water. Hours must pass before their glimmer hardens into the frozen glitter of diamonds. I shall have a long wait before I witness the soundless frolic of the shooting stars. In the profound darkness of certain nights I have seen the sky streaked with so many trailing sparks that it seemed to me a great gale must be blowing through the outer heavens.

Prévot was testing the lamps in their sockets and the emergency torches. Round the bulbs he was wrapping red paper.

"Another layer."

He added another wrapping of paper and touched a swtich. The dim light within the plane was still too bright. As in a photographer's dark-room, it veiled the pale picture of the external world. It hid that glowing phosphorescence which sometimes, at night, clings to the surface of things. Now night has fallen, but it is not yet true night. A crescent moon persists.

Prévot dove aft and came back with a sandwich. I nibbled a bunch of grapes. I was not hungry. I was neither hungry nor thirsty. I felt no weariness. It seemed to me that I could go on like this at the controls for ten years. I was happy.

The moon had set. It was pitch dark when we came in sight of Benghazi. The town lay at the bottom of an obscurity so dense that it was without a halo. I saw the place only when I was over it. As I was hunting for the airdrome the red obstruction lights were switched on. They cut out a black rectangle in the earth.

I banked, and at that moment the rays of a floodlight rose into the sky like a jet from a fire-hose. It pivoted and traced a golden lane over the landing-field. I circled again to get a clear view of what might be in my way. The port was equipped with everything to make a night-landing easy. I throttled down my engine and dropped like a diver into black water.

It was eleven o'clock local time when I landed and taxied across to the beacon. The most helpful ground crew in the world wove in and out of the blinding ray of a searchlight, alternately visible and invisible. They took my papers and began promptly to fill my tanks. Twenty minutes of my time was all they asked for, and I was touched by their great readiness to help. As I was taking off, one of them said:

"Better circle round and fly over us; otherwise we shan't be sure you got off all right."

I rolled down the golden lane towards an unimpeded opening. My *Simoon* lifted her overload clear of the ground well before I reached the end of the runway. The searchlight following me made it hard for me to wheel. Soon it let me go: the men on the ground had guessed that it was dazzling me. I turned right about and banked vertically, and at that moment the searchlight caught me between the eyes again; but scarcely had it touched me when it fled and sent elsewhere its long golden flute. I knew that the ground crew were being most thoughtful and I was grateful. And now I was off to the desert.

All along the line, at Paris, at Tunis, and at Benghazi, I had been told that I should have a following wind of up to twenty-five miles an hour. I was counting on a speed of 190 m.p.h. as I set my course on the middle of the stretch between Alexandria and Cairo. On this course I should avoid the danger zones along the coast, and despite any drifting I might do without knowing it, I should pick up either to port or to starboard the lights of one of those two cities. Failing them I should certainly not miss the lights of the Nile valley. With a steady wind I should reach the Nile in three hours and twenty minutes; if the wind fell, three hours and three-quarters. Calculating thus I began to eat up the six hundred and fifty miles of desert ahead of me.

There was no moon. The world was a bubble of pitch that had dilated until it reached the very stars in the heavens. I should not see a single gleam of light, should not profit by the faintest landmark. Carrying no wireless, I should receive no message from the earth until I reached the Nile. It was useless to try to look at anything other than the compass and the artificial horizon. I might blot the world out of my mind and concentrate my attention upon the slow pulsation of the narrow thread of radium paint that ran along the dark background of the dials.

Whenever Prévot stirred I brought the plane smoothly back to plumb. I went up to six thousand feet where I had been told the winds would be favorable. At long intervals I switched on a lamp to glance at the engine dials, not all of which were phosphorescent;

but most of the time I wrapped myself closely round in darkness among my miniature constellations which gave off the same mineral glow as the stars, the same mysterious and unwearied light, and spoke the same language.

Like the astronomers, I too was reading in the book of celestial mechanics. I too seemed to myself studious and uncorrupted. Everything in the world that might have lured me from my studies had gone out. The external world had ceased to exist.

There was Prévot, who, after a vain resistance, had fallen asleep and left me to the greater enjoyment of my solitude. There was the gentle purr of my beautiful little motor, and before me, on the instrument panel, there were all those tranquil stars. I was most decidedly not sleepy. If this state of quiet well-being persisted until tomorrow night, I intended to push on without a stop to Saïgon.

Now the flight was beginning to seem to me short. Benghazi, the only troublesome night-landing on the route, had banked its fires and settled down behind the horizon in that dark shuttering in which cities take their slumber.

Meanwhile I was turning things over in my mind. We were without the moon's help and we had no wireless. No slightest tenuous tie was to bind us to earth until the Nile showed its thread of light directly ahead of us. We were truly alone in the universe—a thought that caused me not the least worry. If my motor were to cough, that sound would startle me more than if my heart should skip a beat.

Into my mind came the image of Sabathier, the white-haired engineer with the clear eye. I was thinking that, from one point of view, it would be hard to draw a distinction in the matter of human values between a profession like his and that of the painter, the composer, or the poet. I could see in the mind's eye those watchmaker's hands of his that had brought into being this clockwork I was piloting. Men who have given their lives to labors of love go straight to my heart.

"Couldn't I change this?" I had asked him.

"I shouldn't advise it," he had answered.

I was remembering our last conversation. He had thought it inadvisable, and of course that had settled it. A physician, that's it! Exactly the way one puts oneself into the hands of one's doctor—when he has that look in his eye. It was by his motor that we hung suspended in air and were able to go on living with the ticking of time in this penetrable pitch. We were crossing the great dark valley of a fairy-tale, the Valley of Ordeal. Like the prince in the tale, we must meet the test without succor. Failure here would not be forgiven. We were in the lap of the inexorable gods.

A ray of light was filtering through a joint in the lamp shaft. I woke up Prévot and told him to put it out. Prévot stirred in the darkness like a bear, snorted, and came forward. He fumbled for a bit with handkerchiefs and black paper, and the ray of light vanished. That light had bothered me because it was not of my world. It swore at the pale and distant gleam of the phosphorescence and was like a night-club spotlight compared to the gleam of a star. Besides, it had dazzled me and had out-shone all else that gleamed.

We had been flying for three hours. A brightness that seemed to me a glare spurted on the starboard side. I stared. A streamer of light which I had hitherto not noticed was fluttering from a lamp at the tip of the wing. It was an intermittent glow, now brilliant, now dim. It told me that I had flown into a cloud, and it was on the cloud that the lamp was reflected.

I was nearing the landmarks upon which I had counted; a clear sky would have helped a lot. The wing shone bright under the halo. The light steadied itself, became fixed, and then began to radiate in the form of a bouquet of pink blossoms. Great eddies of air were swinging me to and fro. I was navigating somewhere in the belly of a cumulus whose thickness I could not guess. I rose to seventy-five hundred feet and was still in it. Down again to three thousand, and the bouquet of flowers was still with me, motionless and growing brighter.

Well, there it was and there was nothing to do about it. I would think of something else, and wait to get clear of it. Just the same, I did not like this sinister glitter of a one-eyed grog-shop.

"Let me think," I said to myself. "I am bouncing round a bit, but there's nothing abnormal about that. I've been bumped all the way, despite a clear sky and plenty of ceiling. The wind has not died down, and I must be doing better than the 190 m.p.h. I counted on." This was about as far as I could get. Oh, well, when I got through the cloud-bank I would try to take my bearings.

Out of it we flew. The bouquet suddenly vanished, letting me know I was in the clear again. I stared ahead and saw, if one can speak of "seeing" space, a narrow valley of sky and the wall of the next cumulus. Already the bouquet was coming to life again. I was free of that viscous mess from time to time but only for a few seconds each time. After three and a half hours of flying it began to get on my nerves. If I had made the time I imagined, we were certainly approaching the Nile. With a little luck I might be able to spot the river through the rifts, but they were getting rare. I dared not come down, for if I was actually slower than I thought, I was still over high-lying country.

Thus far I was entirely without anxiety; my only fear was that I might presently be wasting time. I decided that I would take things easy until I had flown four and a quarter hours: after that, even in a dead calm (which was highly unlikely) I should have crossed the Nile. When I reached the fringes of the cloud-bank the bouquet winked on and off more and more swiftly and then suddenly went out. Decidedly, I did not like these dot-and-dash messages from the demons of the night.

A green star appeared ahead of me, flashing like a lighthouse. Was it a lighthouse? or really a star? I took no pleasure from this supernatural gleam, this star the Magi might have seen, this dangerous decoy.

Prévot, meanwhile, had waked up and turned his electric torch on the engine dials. I waved him off, him and his torch. We had just sailed into the clear between two clouds and I was busy staring below. Prévot went back to sleep. The gap in the clouds was no help: there was nothing below.

Four hours and five minutes in the air. Prévot awoke and sat down beside me.

"I'll bet we're near Cairo," he said.

"We must be."

"What's that? A star? or is it a lighthouse?"

I had throttled the engine down a little. This, probably, was what had awakened Prévot. He is sensitive to all the variations of sound in flight.

I began a slow descent, intending to slip under the mass of clouds. Meanwhile I had had a look at my map. One thing was sure—the land below me lay at sea level, and there was no risk of conking against a hill. Down I went, flying due north so that the lights of the cities would strike square into my windows. I must have overflown them, and should therefore see them on my left.

Now I was flying below the cumulus. But alongside was another cloud hanging lower down on the left. I swerved so as not to be caught in its net, and headed north-northeast. This second cloud-bank certainly went down a long way, for it blocked my view of the horizon. I dared not give up any more altitude. My altimeter registered 1200 feet, but I had no notion of the atmospheric pressure here. Prévot leaned towards me and I shouted to him, "I'm going out to sea. I'd rather come down on it than risk a crash here."

As a matter of fact, there was nothing to prove that we had not drifted over the sea already. Below that cloud-bank visibility was exactly nil. I hugged my window, trying to read below me, to discover flares, signs of life. I was a man raking dead ashes, trying in vain to retrieve the flame of life in a hearth.

"A lighthouse!"

Both of us spied it at the same moment, that winking decoy! What madness! Where was that phantom light, that invention of the night? For at the very second when Prévot and I leaned forward to pick it out of the air where it had glittered nine hundred feet below our wings, suddenly, at that very instant . . .

"Oh!"

I am quite sure that this was all I said. I am quite sure that all I felt was a terrific crash that rocked our world to its foundations. We had crashed against the earth at a hundred and seventy miles an hour. I am quite sure that in the split second that followed, all I expected was the great flash of ruddy light of the explosion in which Prévot and I were to be blown up together. Neither he nor I had felt the least emotion of any kind. All I could observe in myself was an extraordinary tense feeling of expectancy, the expectancy of that resplendent star in which we were to vanish within the second.

But there was no ruddy star. Instead there was a sort of earthquake that splintered our cabin, ripped away the windows, blew sheets of metal hurtling through space a hundred yards away, and filled our very entrails with its roar. The ship quivered like a knife-blade thrown from a distance into a block of oak, and its anger mashed us as if we were so much pulp.

One second, two seconds passed, and the plane still quivered while I waited with a grotesque impatience for the forces within it to burst it like a bomb. But the subterranean quakings went on without a climax of eruption while I marveled uncomprehendingly at its invisible travail. I was baffled by the quaking, the anger, the interminable postponement. Five seconds passed; six seconds. And suddenly we were seized by a spinning motion, a shock that jerked our cigarettes out of the window, pulverized the starboard wing—and then nothing, nothing but a frozen immobility. I shouted to Prévot:

"Jump!"

And in that instant he cried out:

"Fire!"

We dove together through the wrecked window and found ourselves standing side by side, sixty feet from the plane. I said:

"Are you hurt?"

He answered:

"Not a bit."

But he was rubbing his knee.

"Better run your hands over yourself," I said; "move about a bit. Sure no bones are broken?"

He answered:

"I'm all right. It's that emergency pump."

Emergency pump! I was sure he was going to keel over any minute and split open from head to navel there before my eyes. But he kept repeating with a glassy stare:

"That pump, that emergency pump."

He's out of his head, I thought. He'll start dancing in a minute.

Finally he stopped staring at the plane—which had not gone up in flames—and stared at me instead. And he said again:

"I'm all right. It's that emergency pump. It got me in the knee."

Why we were not blown up, I do not know. I switched on my electric torch and went back over the furrow in the ground traced by the plane. Two hundred and fifty yards from where we stopped the ship had begun to shed the twisted iron and sheet-metal that spattered the sand the length of her traces. We were to see, when day came, that we had run almost tangentially into a gentle slope at the top of a barren plateau. At the point of impact there was a hole in the sand that looked as if it had been made by a plough. Maintaining an even keel, the plane had run its course with the fury and the tail-lashings of a reptile gliding on its belly at the rate of a hundred and seventy miles an hour. We owed our lives to the fact that this desert was surfaced with round black pebbles which had rolled over and over like ball-bearings beneath us. They must have rained upward to the heavens as we shot through them.

Prévot disconnected the batteries for fear of fire by short-circuit. I leaned against the motor and turned the situation over in my mind. I had been flying high for four hours and a quarter, possibly with a thirty-mile following wind. I had been jolted a good deal. If the wind had changed since the weather people forecast it, I was unable to say into what quarter it had veered. All I could make out was that we had crashed in an empty square two hundred and fifty miles on each side.

Prévot came up and sat down beside me.

"I can't believe that we're alive," he said.

I said nothing. Even that thought could not cheer me. A germ of an idea was at work in my mind and was already bothering me. Telling Prévot to switch on his torch as a landmark, I walked straight out, scrutinizing the ground in the light of my own torch as I went.

I went forward slowly, swung round in a wide arc, and changed direction a number of times. I kept my eyes fixed on the ground like a man hunting a lost ring.

Only a little while before I had been straining just as hard to see a gleam of light from the air. Through the darkness I went, bowed over the traveling disk of white light. "Just as I thought," I said to myself, and I went slowly back to the plane. I sat down beside the cabin and ruminated. I had been looking for a reason to hope and had failed to find it. I had been looking for a sign of life, and no sign of life had appeared.

"Prévot, I couldn't find a single blade of grass."

Prévot said nothing, and I was not sure he had understood. Well, we could talk about it again when the curtain rose at dawn. Meanwhile I was dead tired and all I could think was, "Two hundred and fifty miles more or less in the desert."

Suddenly I jumped to my feet. "Water!" I said.

Gas tanks and oil tanks were smashed in. So was our supply of drinking-water. The sand had drunk everything. We found a pint of coffee in a battered thermos flash and half a pint of white wine in another. We filtered both, and poured them into one flask. There were some grapes, too, and a single orange. Meanwhile I was computing: "All this will last us five hours of tramping in the sun."

We crawled into the cabin and waited for dawn. I stretched out, and as I settled down to sleep I took stock of our situation. We didn't know where we were; we had less than a quart of liquid between us; if we were not too far off the Benghazi-Cairo lane we should be found in a week, and that would be too late. Yet it was the best we could hope for. If, on the other hand, we had drifted off our course, we shouldn't be found in six months. One thing was sure—we could not count on being picked up by a plane; the men who came out fot us would have two thousand miles to cover.

"You know, it's a shame," Prévot said suddenly.

"What's a shame?"

"That we didn't crash properly and have it over with."

It seemed pretty early to be throwing in one's hand. Prévot and I pulled ourselves together. There was still a chance, slender as it was, that we might be saved miraculously by a plane. On the other hand, we couldn't stay here and perhaps miss a near-by oasis. We would walk all day and come back to the plane before dark. And before going off we would write our plan in huge letters in the sand.

With this I curled up and settled down to sleep. I was happy to go to sleep. My weariness wrapped me round like a multiple presence. I was not alone in the desert: my drowsiness was peopled with voices and memories and whispered confidences. I was not yet thirsty; I felt strong; and I surrendered myself to sleep as to an aimless journey. Reality lost ground before the advance of dreams.

Ah, but things were different when I awoke!

In times past I have loved the Sahara. I have spent nights alone in the path of marauding tribes and have waked up with untroubled mind in the golden emptiness of the desert where the wind like a sea had raised sandwaves upon its surface. Asleep under the wing of my plane I have looked forward with confidence to being rescued next day. But this was not the Sahara!

Prévot and I walked along the slopes of rolling mounds. The ground was sand covered over with a single layer of shining black pebbles. They gleamed like metal scales and all the domes about us shone like coats of mail. We had dropped down into a mineral world and were hemmed in by iron hills.

When we reached the top of the first crest we saw in the distance another just like it, black and gleaming. As we walked we scraped the ground with our boots, marking a trail over which to return to the plane. We went forward with the sun in our eyes. It was not logical to go due east like this, for everything—the weather reports, the duration of the flight—had made it plain that we had crossed the Nile. But I had started tentatively towards the west and had felt a vague foreboding I could not explain to myself. So I had put off the west till tomorrow. In the same way, provisionally, I had given up going north, though that led to the sea.

Three days later, when scourged by thirst into abandoning the plane and walking straight on until we dropped in our tracks, it was still eastward that we tramped. More precisely, we walked east-northeast. And this too was in defiance of all reason and even of all hope. Yet after we had been rescued we discovered that if we had gone in any other direction we should have been lost.

Northward, we should never have had the endurance to reach the sea. And absurd as it may appear, it seems to me now, since I had no other motive, that I must have chosen the east simply because it was by going eastward that Guillaumet had been saved in the Andes, after I had hunted for him everywhere. In a confused way the east had become for me the direction of life.

We walked on for five hours and then the landscape changed. A river of sand seemed to be running through a valley, and we followed this river bed, taking long strides in order to cover as much ground as possible and get back to the plane before night fell, if our march was in vain. Suddenly I stopped.

"Prévot!"

"What's up?"

"Our tracks!"

How long was it since we had forgotten to leave a wake behind us? We had to find it or die.

We went back, bearing to the right. When we had gone back far enough we would make a right angle to the left and eventually intersect our tracks where we had still remembered to mark them.

This we did and were off again. The heat rose and with it came the mirages. But these were still the commonplace kind—sheets of water that materialized and then vanished as we neared them. We decided to cross the valley of sand and climb the highest dome in order to look round the horizon. This was after six hours of march in which, striding along, we must have covered twenty miles.

When we had struggled up to the top of the black hump we sat down and looked at each other. At our feet lay our valley of sand, opening into a desert of sand whose dazzling brightness seared our eyes. As far as the eye could see lay empty space. But in that space the play of light created mirages which, this time, were of a disturbing kind, fortresses and minarets, angular geometric hulks. I could see also a black mass that pretended to be vegetation, overhung by the last of those clouds that dissolve during the day only to return at night. This mass of vegetation was the shadow of a cumulus.

It was no good going on. The experiment was a failure. We would have to go back to our plane, to that red and white beacon which, perhaps, would be picked out by a flyer. I was not staking great hopes on a rescue party, but it did seem to me our last chance of salvation. In any case, we had to get back to our few drops of liquid, for our throats were parched. We were imprisoned in this iron circle, captives of the curt dictatorship of thirst.

And yet, how hard it was to turn back when there was a chance that we might be on the road to life! Beyond the mirages the horizon was perhaps rich in veritable treasures, in meadows and runnels of sweet water. I knew I was doing the right thing by returning to the plane, and yet as I swung round and started back I was filled with portents of disaster.

We were resting on the ground beside the plane. Nearly forty miles of wandering this day. The last drop of liquid had been drained. No sign of life had appeared to the east. No plane had soared overhead. How long should we be able to hold out? Already our thirst was terrible.

We had built up a great pyre out of bits of the splintered wing. Our gasoline was ready, and we had flung on the heap sheets of metal whose magnesium coating would burn with a hard white flame. We were waiting now for night to come down before we lighted our conflagration. But where were there men to see it?

Night fell and the flames rose. Prayerfully we watched our mute and radiant fanion mount resplendent into the night. As I looked I said to myself that this message was not only a cry for help, it was fraught also with a great deal of love. We were begging water, but we were also begging the communion of human society. Only man can create fire: let another flame light up the night; let man answer man!

I was haunted by a vision of my wife's eyes under the halo of her hat. Of her face I could see only the eyes, questioning me, looking at me yearningly. I am answering, answering with all my strength! What flame could leap higher than this that darts up into the night from my heart?

What I could do, I have done. What we could do, we have done. Nearly forty miles, almost without a drop to drink. Now there was no water left. Was it our fault that we could wait no longer? Suppose we had sat quietly by the plane, taking suck at the mouths of our water-bottles? But from the moment I breathed in the moist bottom of the tin cup, a clock had started up in me.

From the second when I had sucked up the last drop, I had begun to slip downhill. Could I help it if time like a river was carrying me away? Prévot was weeping. I tapped him on the shoulder and said, to console him:

"If we're done for we're done for, and that's all there is to it."

He said:

"Do you think it's me I'm bawling about?"

I might have known it. It was evident enough. Nothing is unbearable. Tomorrow, and the day after, I should learn that nothing was really unbearable. I had never really believed in torture. Reading Poe as a kid, I had already said as much to myself. Once, jammed in the cabin of a plane, I thought I was going to drown; and I had not suffered much. Several times it had seemed to me that the final smash-up was coming, and I don't remember that I thought of it as a cosmic event. And I didn't believe this was going to be agonizing either. There will be time tomorrow to find out stranger things about it. Meanwhile, God knows that despite the bonfire I had decidedly given up hope that our cries would be heard by the world.

"Do you think it's me . . ." There you have what is truly unbearable! Every time I saw those yearning eyes it was as if a flame were searing me. They were like a scream for help, like the flares of a sinking ship. I felt that I should not sit idly by: I should jump up and run—anywhere! straight ahead of me!

What a strange reversal of rôles! But I have always thought it would be like this. Still, I needed Prévot beside me to be quite sure of it. Prévot was a level-headed fellow. He loved life. And yet Prévot no more than I was wringing his hands at the sight of death the way we are told men do. But there did exist something that he could not bear any more than I could. I was perfectly ready to fall asleep, whether for a night or for eternity. If I did fall asleep, I could not even know whether it was for the one or for the other. And the peace of sleep! But that cry that would be sent up at home, that great wail of desolation—that was what I could not bear. I could not stand idly by and look on at that disaster. Each second of silence drove the knife deeper into someone I loved. At the thought, a blind rage surged up in me. Why do these chains bind me and prevent me from rescuing those who are drowning? Why does our conflagration not carry our cry to the ends of the world? Hear me, you out here! Patience. We are coming to save you.

The magnesium had been licked off and the metal was glowing red. There was left only a heap of embers round which we crouched to warm ourselves. Our flaming call had spent itself. Had it set anything in the world in motion? I knew well enough that it hadn't. Here was a prayer that had of necessity gone unheard.

That was that.

I ought to get some sleep.

At daybreak I took a rag and mopped up a little dew on the wings. The mixture of water and paint and oil yielded a spoonful of nauseating liquid which we sipped because it would at least moisten our lips. After this banquet Prévot said:

"Thank God we've got a gun."

Instantly I became furious and turned on him with an aggressiveness which I regretted directly I felt it. There was nothing I should have loathed more at that moment than a gush of sentimentality. I am so made that I have to believe that everything is simple. Birth is simple. Growing up is simple. And dying of thirst is simple. I watched Prévot out of the corner of my eye, ready to wound his feelings, if that was necessary to shut him up.

But Prévot had spoken without emotion. He had been discussing a matter of hygiene, and might have said in the same tone, "We ought to wash our hands." That being so, we were agreed. Indeed already yesterday, my eye falling by chance on the leather holster, the same thought had crossed my mind, and with me too it had been a reasonable reflex, not an emotional one. Pathos resides in social man, not in the individual; what was pathetic was our powerlessness to reassure those for whom we were responsible, not what we might do with the gun.

There was still no sign that we were being sought; or rather they were doubtless hunting for us elsewhere, probably in Arabia. We were to hear no sound of plane until the day after we had abandoned our own. And if ships did pass overhead, what could that mean to us? What could they see in us except two black dots among the thousand shadowy dots in the desert? Absurd to think of being distinguishable from them. None of the reflections that might be attributed to me on the score of this torture would be true. I should not feel in the least tortured. The aerial rescue party would seem to me, each time I sighted one, to be moving through a universe that was not mine. When searchers have to cover two thousand miles of territory, it takes them a good two weeks to spot a plane in the desert from the sky.

They were probably looking for us all along the line from Tripoli to Persia. And still, with all this, I clung to the slim chance that they might pick us out. Was that not our only chance of being saved? I changed my tactics, determining to go reconnoitering by myself. Prévot would get another bonfire together and kindle it in the event that visitors showed up. But we were to have no callers that day.

So off I went without knowing whether or not I should have the stamina to come back. I remembered what I knew about this Libyan desert. When, in the Sahara, humidity is still at forty per cent of saturation, it is only eighteen here in Libya. Life here evaporates like a vapor. Bedouins, explorers, and colonial officers all tell us that a man may go nineteen hours without water. Thereafter his eyes fill with light, and that marks the beginning of the end. The progress made by thirst is swift and terrible. But this northeast wind, this abnormal wind that had blown us out off our course and had marooned us on this plateau, was now prolonging our lives. What was the length of the reprieve it would grant us before our eyes began to fill with light? I went forward with the feeling of a man canoeing in mid-ocean.

I will admit that at daybreak this landscape seemed to me less infernal, and that I began my walk with my hands in my pockets, like a tramp on a highroad. The evening before we had set snares at the mouths of certain mysterious burrows in the ground, and the poacher in me was on the alert. I went first to have a look at our traps. They were empty.

Well, this meant that I should not be drinking blood today; and indeed I hadn't expected to. But though I was not disappointed, my curiosity was aroused. What was there in the desert for these animals to live on? These were certainly the holes of fennecs, a long-eared carnivorous sand-fox the size of a rabbit. I spotted the tracks made by one of them, and gave way to the impulse to follow them. They led to a narrow stream of sand where each footprint was plainly outlined and where I marveled at the pretty palm formed by the three toes spread fanwise on the sand.

I could imagine my little friend trotting blithely along at dawn and licking the dew off the rocks. Here the tracks were wider apart: my fennec had broken into a run. And now I see that a companion has joined him and they have trotted on side by side. These signs of a morning stroll gave me a strange thrill. They were signs of life, and I loved them for that. I almost forgot that I was thirsty.

Finally I came to the pasture-ground of my foxes. Here, every hundred yards or so, I saw sticking up out of the sand a small dry shrub, its twigs heavy with little golden snails. The fennec came here at dawn to do his marketing. And here I was able to observe another of nature's mysteries.

My fennec did not stop at all the shrubs. There were some weighed down with snails which he disdained. Obviously he avoided them with some wariness. Others he stopped at but did not strip of all they bore. He must have picked out two or three shells and then gone on to another restaurant. What was he up to? Was he nurseryman to the snails, encouraging their reproduction by refraining from exhausting the stock on a given shrub, or a given twig? Or was he amusing himself by delaying repletion, putting off satiety in order to enhance the pleasure he took from his morning stroll?

The tracks led me back to the hole in which he lived. Doubtless my fennec crouched below, listening to me and startled by the crunching of my footsteps. I said to him:

"Fox, my little fox, I'm done for; but somehow that doesn't prevent me from taking an interest in your mood."

And there I stayed a bit, ruminating and telling myself that a man was able to adapt himself to anything. The notion that he is to die in thirty years has probably never spoiled any man's fun. Thirty years . . . or thirty days: it's all a matter of perspective.

Only, you have to be able to put certain visions out of your mind.

I went on, finally, and the time came when, along with my

weariness, something in me began to change. If those were not mirages, I was inventing them.

"Hi! Hi, there!"

I shouted and waved my arms, but the man I had seen waving at me turned out to be a black rock. Everything in the desert had grown animate. I stooped to waken a sleeping Bedouin and he turned into the trunk of a black tree. A tree-trunk? Here in the desert? I was amazed and bent over to lift a broken bough. It was solid marble.

Straightening up I looked round and saw more black marble. An antediluvian forest littered the ground with its broken tree-tops. How many thousand years ago, under what hurricane of the time of Genesis, had this cathedral of wood crumbled in this spot? Countless centuries had rolled these fragments of giant pillars at my feet, polished them like steel, petrified and vitrified them and indued them with the color of jet.

I could distinguish the knots in their branches, the twistings of their once living boughs, could count the rings of life in them. This forest had rustled with birds and been filled with music that now was struck by doom and frozen into salt. And all this was hostile to me. Blacker than the chain-mail of the hummocks, these solemn derelicts rejected me. What had I, a living man, to do with this incorruptible stone? Perishable as I was, I whose body was to crumble into dust, what place had I in this eternity?

Since yesterday I had walked nearly fifty miles. This dizziness that I felt came doubtless from my thirst. Or from the sun. It glittered on these hulks until they shone as if smeared with oil. It blazed down on this universal carapace. Sand and fox had no life here. This world was a gigantic anvil upon which the sun beat down. I strode across this anvil and at my temples I could feel the hammer-strokes of the sun.

"Hi! Hi, there!" I called out.

"There is nothing there," I told myself. "Take it easy. You are delirious."

I had to talk to myself aloud, had to bring myself to reason. It was hard for me to reject what I was seeing, hard not to run towards that caravan plodding on the horizon. There! Do you see it?

"Fool! You know very well that you are inventing it."

"You mean that nothing in the world is real?"

Nothing in the world is real if that cross which I see ten miles off on the top of a hill is not real. Or is it a lighthouse? No, the sea does not lie in that direction. Then it must be a cross.

I had spent the night studying my map—but uselessly, since I did not know my position. Still, I had scrutinized all the signs that marked the marvelous presence of man. And somewhere on the map I had seen a little circle surmounted by just such a cross. I had glanced down at the legend to get an explanation of the symbol and had read: "Religious institution."

Close to the cross there had been a black dot. Again I had run my finger down the legend and had read: "Permanent well." My heart had jumped and I had repeated the legend aloud: "Permanent well, permanent well." What were all of Ali Baba's treasures compared with a permanent well? A little farther on were two white circles. "Temporary wells," the legend said. Not quite so exciting. And round about them was nothing . . . unless it was the blankness of despair.

But this must be my "religious institution"! The monks must certainly have planted a great cross on the hill expressly for men in our plight! All I had to do was to walk across to them. I should be taken in by those Dominicans. . . .

"But there are only Coptic monasteries in Libya!" I told myself.

. . . by those learned Dominicans. They have a great cool

kitchen with red tiles, and out in the courtyard a marvelous rusted pump. Beneath the rusted pump; beneath the rusted pump . . . you've guessed it! . . . beneath the rusted pump is dug the permanent well! Ah, what rejoicing when I ring at their gate, when I get my hands on the rope of the great bell.

"Madman! You are describing a house in Provence; and what's more, the house has no bell!"

. . . on the rope of the great bell. The porter will raise his arms to Heaven and cry out, "You are the messenger of the Lord!" and he will call aloud to all the monks. They will pour out of the monastery. They will welcome me with a great feast, as if I were the Prodigal Son. They will lead me to the kitchen and will say to me, "One moment, my son, one moment. We'll just be off to the permanent well." And I shall be trembling with happiness.

No, no! I will *not* weep just because there happens to be no cross on the hill.

The treasures of the west turned out to be mere illusion. I have veered due north. At least the north is filled with the sound of the sea.

Over the hilltop. Look there, at the horizon! The most beautiful city in the world!

"You know perfectly well that is a mirage."

Of course I know it is a mirage! Am I the sort of man who can be fooled? But what if I *want* to go after that mirage? Suppose I enjoy indulging my hope? Suppose it suits me to love that crenelated town all beflagged with sunlight? What if I choose to walk straight ahead on light feet—for you must know that I have dropped my weariness behind me, I am happy now. . . . Prévot and his gun! Don't make me laugh! I prefer my drunkenness. I am drunk. I am dying of thirst.

It took the twilight to sober me. Suddenly I stopped, appalled to think how far I was from our base. In the twilight the mirage was dying. The horizon had stripped itself of its pomp, its palaces, its priestly vestments. It was the old desert horizon again.

"A fine day's work you've done! Night will overtake you. You won't be able to go on before daybreak, and by that time your tracks will have been blown away and you'll be properly nowhere."

In that case I may as well walk straight on. Why turn back? Why should I bring my ship round when I may find the sea straight ahead of me?

"When did you catch a glimpse of the sea? What makes you think you could walk that far? Meanwhile there's Prévot watching for you beside the *Simoon*. He may have been picked up by a caravan, for all you know."

Very good. I'll go back. But first I want to call out for help.

"Hi! Hi!"

By God! You can't tell me this planet is not inhabited. Where are its men?

"Hi! Hi!"

I was hoarse. My voice was gone. I knew it was ridiculous to croak like this, but—one more try:

"Hi! Hi!"

And I turned back.

I had been walking two hours when I saw the flames of the bonfire that Prévot, frightened by my long absence, had sent up. They mattered very little to me now.

Another hour of trudging. Five hundred yards away. A hundred yards. Fifty yards.

"Good Lord!"

Amazement stopped me in my tracks. Joy surged up and filled

my heart with its violence. In the firelight stood Prévot, talking to two Arabs who were leaning against the motor. He had not noticed me, for he was too full of his own joy. If only I had sat still and waited with him! I should have been saved already. Exultantly I called out:

"Hi! Hi!"

The two Bedouins gave a start and stared at me. Prévot left them standing and came forward to meet me. I opened my arms to him. He caught me by the elbow. Did he think I was keeling over? I said:

"At last, eh?"

"What do you mean?"

"The Arabs!"

"What Arabs?"

"Those Arabs there, with you."

Prévot looked at me queerly, and when he spoke I felt as if he was very reluctantly confiding a great secret to me:

"There are no Arabs here."

This time I know I am going to cry.

A man can go nineteen hours without water, and what have we drunk since last night? A few drops of dew at dawn. But the northeast wind is still blowing, still slowing up the process of our evaporation. To it, also, we owe the continued accumulation of high clouds. If only they would drift straight overhead and break into rain! But it never rains in the desert.

"Look here, Prévot. Let's rip up one of the parachutes and spread the sections out on the ground, weighed down with stones. If the wind stays in the same quarter till morning, they'll catch the dew and we can wring them out into one of the tanks."

We spread six triangular sections of parachute under the stars, and Prévot unhooked a fuel tank. This was as much as we could do for ourselves till dawn. But, miracle of miracles! Prévot had come upon an orange while working over the tank. We shared it, and though it was little enough to men who could have used a few gallons of sweet water, still I was overcome with relief.

Stretched out beside the fire I looked at the glowing fruit and said to myself that men did not know what an orange was. "Here we are, condemned to death," I said to myself, "and still the certainty of dying cannot compare with the pleasure I am feeling. The joy I take from this half of an orange which I am holding in my hand is one of the greatest joys I have ever known."

I lay flat on my back, sucking my orange and counting the shooting stars. Here I was, for one minute infinitely happy. "Nobody can know anything of the world in which the individual moves and has his being," I reflected. "There is no guessing it. Only the man locked up in it can know what it is."

For the first time I understood the cigarette and glass of rum that are handed to the criminal about to be executed. I used to think that for a man to accept these wretched gifts at the foot of the gallows was beneath human dignity. Now I was learning that he took pleasure from them. People thought him courageous when he smiled as he smoked or drank. I knew now that he smiled because the taste gave him pleasure. People could not see that his perspective had changed, and that for him the last hour of his life was a life in itself.

We collected an enormous quantity of water—perhaps as much as two quarts. Never again would we be thirsty! We were saved; we had a liquid to drink!

I dipped my tin cup into the tank and brought up a beautifully yellow-green liquid the first mouthful of which nauseated me so that despite my thirst I had to catch my breath before swallowing it. I would have swallowed mud, I swear; but this taste of poisonous metal cut keener than thirst.

I glanced at Prévot and saw him going round and round with his eyes fixed to the ground as if looking for something. Suddenly he leaned forward and began to vomit without interrupting his spinning. Half a minute later it was my turn. I was seized by such convulsions that I went down on my knees and dug my fingers into the sand while I puked. Neither of us spoke, and for a quarter of an hour we remained thus shaken, bringing up nothing but a little bile.

After a time it passed and all I felt was a vague, distant nausea. But our last hope had fled. Whether our bad luck was due to a sizing on the parachute or to the magnesium lining of the tank, I never found out. Certain it was that we needed either another set of cloths or another receptacle.

Well, it was broad daylight and time we were on our way. This time we should strike out as fast as we could, leave this cursed plateau, and tramp till we dropped in our tracks. That was what Guillaumet had done in the Andes. I had been thinking of him all the day before and had determined to follow his example. I should do violence to the pilot's unwritten law, which is to stick by the ship; but I was sure no one would be along to look for us here.

Once again we discovered that it was not we who were shipwrecked, not we but those who were waiting for news of us, those who were alarmed by our silence, were already torn with grief by some atrocious and fantastic report. We could not but strive towards them. Guillaumet had done it, had scrambled towards his lost ones. To do so is a universal impulse.

"If I were alone in the world," Prévot said, "I'd lie down right here. Damned if I wouldn't."

East-northeast we tramped. If we had in fact crossed the Nile, each step was leading us deeper and deeper into the desert.

I don't remember anything about that day. I remember only my haste. I was hurrying desperately towards something—towards some finality. I remember also that I walked with my eyes to the ground, for the mirages were more than I could bear. From time to time we would correct our course by the compass, and now and again we would lie down to catch our breath. I remember having flung away my waterproof, which I had held on to as covering for the night. That is as much as I recall about the day. Of what happened when the chill of evening came, I remember more. But during the day I had simply turned to sand and was a being without mind.

When the sun set we decided to make camp. Oh, I knew as well as anybody that we should push on, that this one waterless night would finish us off. But we had brought along the bits of parachute, and if the poison was not in the sizing, we might get a sip of water next morning. Once again we spread our trap for the dew under the stars.

But the sky in the north was cloudless. The wind no longer had the same taste on the lip. It had moved into another quarter. Something was rustling against us, but this time it seemed to be the desert itself. The wild beast was stalking us, had us in its power. I could feel its breath in my face, could feel it lick my face and hands. Suppose I walked on: at the best I could do five or six miles more. Remember that in three days I had covered one hundred miles, practically without water.

And then, just as we stopped, Prévot said:

"I swear to you I see a lake!"

"You're crazy."

"Have you ever heard of a mirage after sunset?" he challenged.

I didn't seem able to answer him. I had long ago given up believing my own eyes. Perhaps it was not a mirage; but in that case it was a hallucination. How could Prévot go on believing? But he was stubborn about it.

"It's only twenty minutes off. I'll go have a look."

His mulishness got on my nerves.

"Go ahead!" I shouted. "Take your little constitutional. Nothing better for a man. But let me tell you, if your lake exists it is salt. And whether it's salt or not, it's a devil of a way off. And besides, there is no damned lake!"

Prévot was already on his way, his eyes glassy. I knew the strength of these irresistible obsessions. I was thinking: "There are somnambulists who walk straight into locomotives." And I knew that Prévot would not come back. He would be seized by the vertigo of empty space and would be unable to turn back. And then he would keel over. He somewhere, and I somewhere else. Not that it was important.

Thinking thus, it struck me that this mood of resignation was doing me no good. Once when I was half drowned I had let myself go like this. Lying now flat on my face on the stony ground, I took this occasion to write a letter for posthumous delivery. It gave me a chance, also, to take stock of myself again. I tried to bring up a little saliva: how long was it since I had spit? No saliva. If I kept my mouth closed, a kind of glue sealed my lips together. It dried on the outside of the lips and formed a hard crust. However, I found I was still able to swallow, and I bethought me that I was still not seeing a blinding light in my eyes. Once I was treated to that radiant spectacle I might know that the end was a couple of hours away.

Night fell. The moon had swollen since I last saw it. Prévot was still not back. I stretched out on my back and turned these few data over in my mind. A familiar impression came over me, and I tried to seize it. I was . . . I was . . . I was at sea. I was on a ship going to South America and was stretched out, exactly like this, on the boat deck. The tip of the mast was swaying to and fro, very slowly, among the stars. That mast was missing tonight, but again I was at sea, bound for a port I was to make without raising a finger. Slave-traders had flung me on this ship.

I thought of Prévot who was still not back. Not once had I heard him complain. That was a good thing. To hear him whine would have been unbearable. Prévot was a man.

What was that! Five hundred yards ahead of me I could see the light of his lamp. He had lost his way. I had no lamp with which to signal back. I stood up and shouted, but he could not hear me.

A second lamp, and then a third! God in Heaven! It was a search party and it was me they were hunting!

"Hi! Hi!" I shouted.

But they had not heard me. The three lamps were still signaling me.

"Tonight I am sane," I said to myself. "I am relaxed. I am not out of my head. Those are certainly three lamps and they are about five hundred yards off." I stared at them and shouted again, and again I gathered that they could not hear me.

Then, for the first and only time, I was really seized with panic. I could still run, I thought. "Wait! Wait!" I screamed. They seemed to be turning away from me, going off, hunting me elsewhere! And I stood tottering, tottering on the brink of life when there were arms out there ready to catch me! I shouted and screamed again and again.

They had heard me! An answering shout had come. I was

strangling, suffocating, but I ran on, shouting as I ran, until I saw Prévot and keeled over.

When I could speak again I said: "Whew! When I saw all those light . . ."

"What lights?"

God in Heaven, it was true! He was alone!

This time I was beyond despair. I was filled with a sort of dumb fury.

"What about your lake?" I rasped.

"As fast as I moved towards it, it moved back. I walked after it for about half an hour. Then it seemed still too far away, so I came back. But I am positive, now, that it is a lake."

"You're crazy. Absolutely crazy. Why did you do it? Tell me. Why?"

What had be done? Why had he done it? I was ready to weep with indignation, yet I scarcely knew why I was so indignant. Prévot mumbled his excuse:

"I felt I had to find some water. You . . . your lips were awfully pale."

Well! My anger died within me. I passed my hand over my forehead as if I were waking out of sleep. I was suddenly sad. I said:

"There was no mistake about it. I saw them as clearly as I see you now. Three lights there were. I tell you, Prévot, I saw them!"

Prévot made no comment.

"Well," he said finally, "I guess we're in a bad way."

In this air devoid of moisture the soil is swift to give off its temperature. It was already very cold. I stood up and stamped about. But soon a violent fit of trembling came over me. My dehydrated blood was moving sluggishly and I was pierced by a freezing chill which was not merely the chill of night. My teeth were chattering and my whole body had begun to twitch. My hand shook so that I could not hold an electric torch. I who had never been sensitive to cold was about to die of cold. What a strange effect thirst can have!

Somewhere, tired of carrying it in the sun, I had let my waterproof drop. Now the wind was growing bitter and I was learning that in the desert there is no place of refuge. The desert is as smooth as marble. By day it throws no shadows; by night it hands you over naked to the wind. Not a tree, not a hedge, not a rock behind which I could seek shelter. The wind was charging me like a troop of cavalry across open county. I turned and twisted to escape it: I lay down, stood up, lay down again, and still I was exposed to its freezing lash. I had no strength to run from the assassin and under the sabre-stroke I tumbled to my knees, my head between my hands.

A little later I pieced these bits together and remembered that I had struggled to my feet and had started to walk on, shivering as I went. I had started forward wondering where I was and then I had heard Prévot. His shouting had jolted me into consciousness.

I went back towards him, still trembling from head to foot—quivering with the attack of hiccups that was convulsing my whole body. To myself I said: "It isn't the cold. It's something else. It's the end." The simple fact was that I hadn't enough water in me. I had tramped too far yesterday and the day before when I was off by myself, and I was dehydrated.

The thought of dying of the cold hurt me. I preferred the phantoms of my mind, the cross, the trees, the lamps. At least they would have killed me by enchantment. But to be whipped to death like a slave! . . .

Confound it! Down on my knees again! We had with us a little store of medicines—a hundred grammes of ninety per cent alcohol, the same of pure ether, and a small bottle of iodine. I tried to swallow a little of the ether: it was like swallowing a knife. Then I tried the alcohol: it contracted my gullet. I dug a pit in the sand, lay down in it, and flung handfuls of sand over me until all but my face was buried in it.

Prévot was able to collect a few twigs, and he lit a fire which soon burnt itself out. He wouldn't bury himself in the sand, but preferred to stamp round and round in a circle. That was foolish.

My throat stayed shut, and though I knew that was a bad sign, I felt better. I felt calm. I felt a peace that was beyond all hope. Once more, despite myself, I was journeying, trussed up on the deck of my slave-ship under the stars. It seemed to me that I was perhaps not in such a bad pass after all.

So long as I lay absolutely motionless, I no longer felt the cold. This allowed me to forget my body buried in the sand. I said to myself that I would not budge an inch, and would therefore never suffer again. As a matter of fact, we really suffer very little. Back of all these torments there is the orchestration of fatigue or of delirium, and we live on in a kind of picture-book, a slightly cruel fairy-tale.

A little while ago the wind had been after me with whip and spur, and I was running in circles like a frightened fox. After that came a time when I couldn't breathe. A great knee was crushing my chest. A knee. I was writhing in vain to free myself from the weight of the angel who had overthrown me. There had not been a moment when I was alone in this desert. But now I have ceased to believe in my surroundings; I have withdrawn into myself, have shut my eyes, have not so much as batted an eyelid. I have the feeling that this torrent of visions is sweeping me away to a tranquil dream: so rivers cease their turbulence in the embrace of the sea.

Farewell, eyes that I loved! Do not blame me if the human body cannot go three days without water. I should never have believed that man was so truly the prisoner of the springs and freshets. I had no notion that our self-sufficiency was so circumscribed. We take it for granted that a man is able to stride straight out into the world. We believe that man is free. We never see the cord that binds him to wells and fountains, that umbilical cord by which he is tied to the womb of the world. Let man take but one step too many . . . and the cord snaps.

Apart from your suffering, I have no regrets. All in all, it has been a good life. If I got free of this I should start right in again. A man cannot live a decent life in cities, and I need to feel myself live. I am not thinking of aviation. The airplane is a means, not an end. One doesn't risk one's life for a plane any more than a farmer ploughs for the sake of the plough. But the airplane is a means of getting away from towns and their bookkeeping and coming to grips with reality.

Flying is a man's job and its worries are a man's worries. A pilot's business is with the wind, with the stars, with night, with sand, with the sea. He strives to outwit the forces of nature. He stares in expectancy for the coming of dawn the way a gardener awaits the coming of spring. He looks forward to port as to a promised land, and truth for him is what lives in the stars.

I have nothing to complain of. For three days I have tramped the desert, have known the pangs of thirst, have followed false scents in the sand, have pinned my faith on the dew. I have struggled to rejoin my kind, whose very existence on earth I had forgotten. These are the cares of men alive in every fibre, and I cannot help thinking them more important than the fretful choosing of a night-club in which to spend the evening. Compare the one life with the other, and all things considered this is luxury! I have no regrets.

I have gambled and lost. It was all in the day's work. At least I have had the unforgettable taste of the sea on my lips.

I am not talking about living dangerously. Such words are meaningless to me. The toreador does not stir me to enthusiasm. It is not danger I love. I know what I love. It is life.

The sky seemed to me faintly bright. I drew up one arm through the sand. There was a bit of the torn parachute within reach, and I ran my hand over it. It was bone dry. Let's see. Dew falls at dawn. Here was dawn risen and no moisture on the cloth. My mind was befuddled and I heard myself say: "There is a dry heart here, a dry heart that cannot know the relief of tears."

I scrambled to my feet. "We're off, Prévot," I said. "Our throats are still open. Get along, man!"

The wind that shrivels up a man in nineteen hours was now blowing out of the west. My gullet was not yet shut, but it was hard and painful and I could feel that there was a rasp in it. Soon that cough would begin that I had been told about and was now expecting. My tongue was becoming a nuisance. But most serious of all, I was beginning to see shining spots before my eyes. When those spots changed into flames, I should simply lie down.

The first morning hours were cool and we took advantage of them to get on at a good pace. We knew that once the sun was high there would be no more walking for us. We no longer had the right to sweat. Certainly not to stop and catch our breath. This coolness was merely the coolness of low humidity. The prevailing wind was coming from the desert, and under its soft and treacherous caress the blood was being dried out of us.

Our first day's nourishment had been a few grapes. In the next three days each of us ate half an orange and a bit of cake. If we had had anything left now, we couldn't have eaten it because we had no saliva with which to masticate it. But I had stopped being hungry. Thirsty I was, yes, and it seemed to me that I was suffering less from thirst itself than from the effects of thirst. Gullet hard. Tongue like plaster-of-Paris. A rasping in the throat. A horrible taste in the mouth.

All these sensations were new to me, and though I believed water could rid me of them, nothing in my memory associated them with water. Thirst had become more and more a disease and less and less a craving. I began to realize that the thought of water and fruit was now less agonizing than it had been. I was forgetting the radiance of the orange, just as I was forgetting the eyes under the hat-brim. Perhaps I was forgetting everything.

We had sat down after all, but it could not be for long. Nevertheless, it was impossible to go five hundred yards without our legs giving way. To stretch out on the sand would be marvelous—but it could not be.

The landscape had begun to change. Rocky places grew rarer and the sand was now firm beneath our feet. A mile ahead stood dunes and on those dunes we could see a scrubby vegetation. At least this sand was preferable to the steely surface over which we had been trudging. This was the golden desert. This might have been the Sahara. It was in a sense my country.

Two hundred yards had now become our limit, but we had determined to carry on until we reached the vegetation. Better than that we could not hope to do. A week later, when we went back over our traces in a car to have a look at the *Simoon,* I measured this last lap and found that it was just short of fifty miles. All told we had done one hundred and twenty-four miles.

The previous day I had tramped without hope. Today the word "hope" had grown meaningless. Today we were tramping simply because we were tramping. Probably oxen work for the same reason. Yesterday I had dreamed of a paradise of orange-trees. Today I would not give a button for paradise; I did not believe oranges existed. When I thought about myself I found in me nothing but a heart squeezed dry. I was tottering but emotionless. I felt no distress whatever, and in a way I regretted it: misery would have seemed to me as sweet as water. I might then have felt sorry for myself and commiserated with myself as with a friend. But I had not a friend left on earth.

Later, when we were rescued, seeing our burnt-out eyes men thought we must have called aloud and wept and suffered. But cries of despair, misery, sobbing grief are a kind of wealth, and we possessed no wealth. When a young girl is disappointed in love she weeps and knows sorrow. Sorrow is one of the vibrations that prove the fact of living. I felt no sorrow. I was the desert. I could no longer bring up a little saliva; neither could I any longer summon those moving visions towards which I should have loved to stretch forth arms. The sun had dried up the springs of tears in me.

And yet, what was that? A ripple of hope went through me like a faint breeze over a lake. What was this sign that had awakened my instinct before knocking on the door of my consciousness? Nothing had changed, and yet everything was changed. This sheet of sand, these low hummocks and sparse tufts of verdure that had been a landscape, were now become a stage setting. Thus far the stage was empty, but the scene was set. I looked at Prévot. The same astonishing thing had happened to him as to me, but he was as far from guessing its significance as I was.

I swear to you that something is about to happen. I swear that life has sprung in this desert. I swear that this emptiness, this stillness, has suddenly become more stirring than a tumult on a public square.

"Prévot! Footprints! We are saved!"

We had wandered from the trail of the human species; we had cast ourselves forth from the tribe; we had found ourselves alone on earth and forgotten by the universal migration; and here, imprinted in the sand, were the divine and naked feet of man!

"Look, Prévot, here two men stood together and then separated."

"Here a camel knelt."

"Here . . ."

But it was not true that we were already saved. It was not enough to squat down and wait. Before long we should be past saving. Once the cough had begun, the progress made by thirst is swift.

Still, I believed in that caravan swaying somewhere in the desert, heavy with its cargo of treasure.

We went on. Suddenly I heard a cock crow. I remembered what Guillaumet had told me: "Towards the end I heard cocks crowing in the Andes. And I heard the railway train." The instant the cock crowed I thought of Guillaumet and I said to myself: "First it was my eyes that played tricks on me. I suppose this is another of the effects of thirst. Probably my ears have merely held out longer than my eyes." But Prévot grabbed my arm:

"Did you hear that?"

"What?"

"The cock."

"Why . . . why, yes, I did."

To myself I said: "Fool! Get it through your head! This means life!"

I had one last hallucination—three dogs chasing one another. Prévot looked, but could not see them. However, both of us waved our arms at a Bedouin. Both of us shouted with all the breath in our bodies, and laughed for happiness.

But our voices could not carry thirty yards. The Bedouin on his slow-moving camel had come into view from behind a dune and now he was moving slowly out of sight. The man was probably the only Arab in this desert, sent by a demon to materialize and vanish before the eyes of us who could not run.

We saw in profile on the dune another Arab. We shouted, but our shouts were whispers. We waved our arms and it seemed to us that they must fill the sky with monstrous signals. Still the Bedouin stared with averted face away from us.

At last, slowly, slowly he began a right angle turn in our direction. At the very second when he came face to face with us, I thought, the curtain would come down. At the very second when his eyes met ours, thirst would vanish and by this man would death and the mirages be wiped out. Let this man but make a quarter-turn left and the world is changed. Let him but bring his torso round, but sweep the scene with a glance, and like a god he can create life.

The miracle had come to pass. He was walking towards us over the sand like a god over the waves.

The Arab looked at us without a word. He placed his hands upon our shoulders and we obeyed him: we stretched out upon the sand. Race, language, religion were forgotten. There was only this humble nomad with the hands of an archangel on our shoulders.

Face to the sand, we waited. And when the water came, we drank like calves with our faces in the basin, and with a greediness which alarmed the Bedouin so that from time to time he pulled us back. But as soon as his hand fell away from us we plunged our faces anew into the water.

Water, thou hast no taste, no color, no odor; canst not be defined, art relished while ever mysterious. Not necessary to life, but rather life itself, thou fillest us with a gratification that exceeds the delight of the senses. By thy might, there return into us treasures that we had abandoned. By thy grace, there are released in us all the dried-up runnels of our heart. Of the riches that exist in the world, thou art the rarest and also the most delicate—thou so pure within the bowels of the earth! A man may die of thirst lying beside a magnesian spring. He may die within reach of a salt lake. He may die though he hold in his hand a jug of dew, if it be inhabited by evil salts. For thou, water, art a proud divinity, allowing no alteration, no foreignness in thy being. And the joy that thou spreadest is an infinitely simple joy.

You, Bedouin of Libya who saved our lives, though you will dwell for ever in my memory yet I shall never be able to recapture your features. You are Humanity and your face comes into my mind simply as man incarnate. You, our beloved fellowman, did not know who we might be, and yet you recognized us without fail. And I, in my turn, shall recognize you in the faces of all mankind. You came towards me in an aureole of charity and magnanimity bearing the gift of water. All my friends and all my enemies marched towards me in your person. It did not seem to me that you were rescuing me: rather did it seem that you were forgiving me. And I felt I had no enemy left in all the world.

This is the end of my story. Lifted on a camel, we went on for three hours. Then, broken with weariness, we asked to be set down at a camp while the cameleers went on ahead for help. Towards six in the evening a car manned by armed Bedouins came to fetch us. A half-hour later we were set down at the house of a Swiss engineer named Raccaud who was operating a soda factory beside saline deposits in the desert. He was unforgettably kind to us. By midnight we were in Cairo.

I awoke between white sheets. Through the curtains came the rays of a sun that was no longer an enemy. I spread butter and honey on my bread. I smiled. I recaptured the savor of my childhood and all its marvels. And I read and re-read the telegram from those dearest to me in all the world whose three words had shattered me:

"So terribly happy!"

# IX

## *Barcelona and Madrid (1936)*

Once again I had found myself in the presence of a truth and had failed to recognize it. Consider what had happened to me: I had thought myself lost, had touched the very bottom of despair; and then, when the spirit of renunciation had filled me, I had known peace. I know now what I was not conscious of at the time—that in such an hour a man feels that he has finally found himself and has become his own friend. An essential inner need has been satisfied, and against that satisfaction, that self-fulfilment, no external power can prevail. Bonnafous, I imagine, he who spent his life racing before the wind, was acquainted with this serenity of spirit. Guillaumet, too, in his snows. Never shall I forget that, lying buried to the chin in sand, strangled slowly to death by thirst, my heart was infinitely warm beneath the desert stars.

What can men do to make known to themselves this sense of deliverance? Everything about mankind is paradox. He who strives and conquers grows soft. The magnanimous man grown rich becomes mean. The creative artist for whom everything is made easy nods. Every doctrine swears that it can breed men, but none can tell us in advance what sort of men it will breed. Men are not cattle to be fattened for market. In the scales of life an indigent Newton weighs more than a parcel of prosperous nonentities. All of us have had the experience of a sudden joy that came when nothing in the world had forewarned us of its coming—a joy so thrilling that if it was born of misery we remembered even the misery with tenderness. All of us, on seeing old friends again, have remembered with happiness the trials we lived through with those friends. Of what can we be certain except this—that we are fertilized by mysterious circumstances? Where is man's truth to be found?

Truth is not that which can be demonstrated by the aid of logic. If orange-trees are hardy and rich in fruit in this bit of soil and not that, then this bit of soil is what is truth for orange-trees. If a particular religion, or culture, or scale of values, if one form of activity rather than another, brings self-fulfilment to a man, releases the prince asleep within him unknown to himself, then that scale of values, that culture, that form of activity, constitute his truth. Logic, you say? Let logic wangle its own explanation of life.

Because it is man and not flying that concerns me most, I shall close this book with the story of man's gropings towards self-fulfilment as I witnessed them in the early months of the civil war in Spain. One year after crashing in the desert I made a tour of the Catalan front in order to learn what happens to man when the scaffolding of his traditions suddenly collapses. To Madrid I went for an answer to another question: How does it happen that men are sometimes willing to die?

I

Flying west from Lyon, I veered left in the direction of the Pyrenees and Spain. Below me floated fleecy white clouds, summer clouds, clouds made for amateur flyers in which great gaps opened like skylights. Through one of these windows I could see Perpignan lying at the bottom of a well of light.

I was flying solo, and as I looked down on Perpignan I was day-dreaming. I had spent six months there once while serving as test pilot at a near-by airdrome. When the day's work was done I would drive into this town where every day was as peaceful as Sunday. I would sit in a wicker chair within sound of the café band, sip a glass of port, and look idly on at the provincial life of the place, reflecting that it was as innocent as a review of lead soldiers. These pretty girls, these carefree strollers, this pure sky. . . .

But here came the Pyrenees. The last happy town was left behind.

Below me lay Figueras, and Spain. This was where men killed one another. What was most astonishing here was not the sight of conflagration, ruin, and signs of man's distress—it was the absence of all these. Figueras seemed no different from Perpignan. I leaned out and stared hard.

There were no scars on that heap of white gravel, that church gleaming in the sun, which I knew had been burnt. I could not distinguish its irreparable wounds. Gone was the pale smoke that had carried off its gilding, had melted in the blue of the sky its altar screens, its prayer books, its sacerdotal treasures. Not a line of the church was altered. This town, seated at the heart of its fan-shaped roads like a spider at the centre of its silken trap, looked very much like the other.

Like other towns, this one was nourished by the fruits of the plain that rose along the white highways to meet it. All that I could discern was the slow gnawing which, through the centuries, had swallowed up the soil, driven away the forests, divided up the fields, dug out these life-giving irrigation ditches. Here was a face unlikely to change much, for it was already old. A colony of bees, I said to myself, once it was established so solidly within the boundaries of an acre of flowers, would be assured of peace. But peace is not given to a colony of men.

Human drama does not show itself on the surface of life. It is not played out in the visible world, but in the hearts of men. Even in happy Perpignan a victim of cancer walled up behind his hospital window goes round and round in a circle striving helplessly to escape the pain that hovers over him like a relentless kite. One man in misery can disrupt the peace of a city. It is another of the miraculous things about mankind that there is no pain nor passion that does not radiate to the ends of the earth. Let a man in a garret but burn with enough intensity and he will set fire to the world.

Gerona went by, Barcelona loomed into view, and I let myself glide gently down from the perch of my observatory. Even here I could see nothing out of the way, unless it was that the avenues were deserted. Again there were devastated churches which, from above, looked untouched. Faintly visible was something that I guessed to be smoke. Was that one of the signs I was seeking? Was this a scrap of evidence of that nearly soundless anger whose all-destroying wrath was so hard to measure? A whole civilization was contained in that faint golden puff so lightly dispersed by a breath of wind.

I am quite convinced of the sincerity of people who say: "Terror in Barcelona? Nonsense. That great city in ashes? A mere twenty houses wrecked. Streets heaped with the dead? A few hundred killed out of a population of a million. Where did you see a firing line running with blood and deafening with the roar of guns?"

I agree that I saw no firing line. I saw groups of tranquil men and women strolling on the Ramblas. When, on occasion, I ran against a barricade of militiamen in arms, a smile was often enough to open the way before me. I did not come at once upon the firing line. In a civil war the firing line is invisible; it passes through the hearts of men. And yet, on my very first night in Barcelona I skirted it.

I was sitting on the pavement of a café, sipping my drink surrounded by light-hearted men and women, when suddenly four armed men stopped where I sat, stared at a man at the next table, and without a word pointed their guns at his stomach. Streaming with sweat the man stood up and raised leaden arms above his head. One of the militiamen ran his hands over his clothes and his eyes over some papers he found in the man's pockets, and ordered him to come along.

The man left his half-emptied glass, the last glass of his life, and started down the road. Surrounded by the squad, his hands stuck up like the hands of a man going down for the last time.

"Fascist!" A woman behind me said it with contempt. She was the only witness who dared betray that anything out of the ordinary had taken place. Untouched, the man's glass stood on the table, a mute witness to a mad confidence in chance, in forgiveness, in life. I sat watching the disappearance in a ring of rifles of a man who five minutes before, within two feet of me, had crossed the invisible firing line.

My guides were anarchists. They led me to the railway station where troops were being entrained. Far from the platforms built for tender farewells, we were walking in a desert of signal towers and switching points, stumbling in the rain through a labyrinthine yard filled with blackened goods wagons where tarpaulins the color of lard were spread over carloads of stiffened forms. This world had lost its human quality, had become a world of iron, and therefore uninhabitable. A ship remains a living thing only so long as man with his brushes and oils swabs an artificial layer of light over it. Leave them to themselves a couple of weeks and the life dies out of your ship, your factory, your railway; death covers their faces. After six thousand years the stones of a temple still vibrate with the passage of man; but a little rust, a night of rain, and this railway yard is eaten away to its very skeleton.

Here are our men. Cannon and machine-guns are being loaded on board with the straining muscles and the hoarse gaspings that are always drawn from men by these monstrous insects, these fleshless insects, these lumps of carapace and vertebra. What is startling here is the silence. Not a note of song, not a single shout. Only, now and then, when a gun-carriage lands, the hollow thump of a steel plate. Of human voices no sound.

No uniforms, either. These men are going off to be killed in their working garb. Wearing their dark clothes stiff with mud, the column heaving and sweating at their work look like the denizens of a night shelter. They fill me with the same uneasiness I felt when the yellow fever broke out among us at Dakar, ten years ago.

The chief of the detachment had been speaking to me in a whisper. I caught the end of his speech:

". . . and we move up to Saragossa."

Why the devil did he have to whisper! The atmosphere of this yard made me think of a hospital. But of course! That was it. A civil war is not a war, it is a disease. These men were not going up to the front in the exultation of certain victory; they were struggling blindly against infection.

And the same thing was going on in the enemy camp. The purpose of this struggle was not to rid the country of an invading foreigner but to eradicate a plague. A new faith is like a plague. It attacks from within. It propagates in the invisible. Walking in the streets, whoever belongs to a Party feels himself surrounded by secretly infected men.

This must have been why these troops were going off in silence with their instruments of asphyxiation. There was not the slightest resemblance between them and regiments that go into battle against foreign armies and are set out on the chessboard of the fields and moved about by strategists. These men had gathered together haphazardly in a city filled with chaos.

There was not much to choose between Barcelona and its enemy, Saragossa: both were composed of the same swarm of communists, anarchists, and fascists. The very men who collected on the same side were perhaps more different from one another than from their enemies. In civil war the enemy is inward; one as good as fights against oneself.

What else can explain the particular horror of this war in which firing squads count for more than soldiers of the line? Death in this war is a sort of quarantine. Purges take place of germ-carriers. The anarchists go from house to house and load the plague-stricken into their tumbrils, while on the other side of the barricade Franco is able to utter that horrible boast: "There are no more communists among us."

The conscripts are weeded out by a kind of medical board; the officer in charge is a sort of army doctor. Men present themselves for service with pride shining in their eyes and the belief in their hearts that they have a part to play in society.

"Exempt from service for life!" is the decision.

Fields have been turned into charnel-houses and the dead are burned in lime or petroleum. Respect for the dignity of man has been trampled under foot. Since on both sides the political parties spy upon the stirrings of man's conscience as upon the workings of a disease, why should the urn of his flesh be respected? This body that clothes the spirit, that moves with grace and boldness, that knows love, that is apt for self-sacrifice—no one now so much as thinks of giving it decent burial.

I thought of our respect for the dead. I thought of the white sanatorium where the light of a man's life goes quietly out in the presence of those who love him and who garner as if it were an inestimable treasure his last words, his ultimate smile. How right they are! Seeing that this same whole is never again to take shape in the world. Never again will be heard exactly that note of laughter, that intonation of voice, that quality of repartee. Each individual is a miracle. No wonder we go on speaking of the dead for twenty years.

Here, in Spain, a man is simply stood up against a wall and he gives up his entrails to the stones of the courtyard. You have been captured. You are shot. Reason: your ideas were not our ideas.

This entrainment in the rain is the only thing that rings true about their war. These men stand round and stare at me, and I read in their eyes a mournful sobriety. They know the fate that awaits them if they are captured. I begin to shiver with the cold and observe of a sudden that no woman has been allowed to see them off.

The absence of women seems to me right. There is no place here for mothers who bring children into the world in ignorance of the faith that will some day flare up in their sons, in ignorance of the ideologist who, according to his lights, will prop up their sons against a wall when they have come to their twenty years of life.

We went up by motor into the war zone. Barricades became more frequent, and from place to place we had to negotiate with revolutionary committees. Passes were valid only from one village to the next.

"Are you trying to get closer to the front?"

"Exactly."

The chairman of the local committee consulted a large-scale map.

"You won't be able to get through. The rebels have occupied the road four miles ahead. But you might try swinging left here. This road ought to be free. Though there was talk of rebel cavalry cutting it this morning."

It was very difficult in those early days of the revolution to know one's way about in the vicinity of the front. There were loyal villages, rebel villages, neutral villages, and they shifted their allegiance between dawn and dark. This tangle of loyal and rebel zones made me think the push must be pretty weak. It certainly bore no resemblance to a line of trenches cutting off friend from enemy as cleanly as a knife. I felt as if I were walking in a bog. Here the earth was solid beneath our feet: there we sank into it. We moved in a maze of uncertainty. Yet what space, what air between movements! These military operations are curiously lacking in density.

Once again we reached a point beyond which we were told we could not advance. Six rifles and a low wall of paving stones blocked the road. Four men and two women lay stretched on the ground behind the wall. I made a mental note that the women did not know how to hold a rifle.

"This is as far as you can go."

"Why?"

"Rebels."

We got out of the car and sat down with the militiamen upon the grass. They put down their rifles and cut a few slices of fresh bread.

"Is this your village?" we asked.

"No, we are Catalans, from Barcelona. Communist Party."

One of the girls stretched herself and sat up on the barricade, her hair blowing in the wind. She was rather thick-set, but young and healthy. Smiling happily she said:

"I am going to stay in this village when the war is over. I didn't know it, but the country beats the city all hollow."

She cast a loving glance round at the countryside, as if stirred by a revelation. Her life had been the gray slums, days spent in a factory, and the sordid compensation afforded by the cafés. Everything that went on here seemed to her as jolly as a picnic. She jumped down and ran to the village well. Probably she believed she was drinking at the very breast of mother earth.

"Have you done any fighting here?"

"No. The rebels kick up a little dust now and then, but . . . We see a lorryload of men from time to time and hope that they will come along this road. But nothing has come by in two weeks."

They were awaiting their first enemy. In the rebel village opposite sat another half-dozen militiamen awaiting a first enemy. Twelve warriors alone in the world.

Each side was waiting for something to be born in the invisible. The rebels were waiting for the host of hesitant people in Madrid to declare themselves for Franco. Barcelona was waiting for Saragossa to waken out of an inspired dream, declare itself Socialist, and fall. It was the thought more than the soldier that was besieging the town. The thought was the great hope and the great enemy.

It seemed to me that the bombers, the shells, the militiamen under arms, by themselves had no power to conquer. On each side a single man entrenched behind his line of defense was better than a hundred besiegers. But thought might worm its way in.

From time to time there is an attack. From time to time the tree is shaken. Not to uproot it, but merely to see if the fruit is yet ripe. And if it is, a town falls.

## II

Back from the front, I found friends in Barcelona who allowed me to join in their mysterious expeditions. We went deep into the mountains and were now in one of those villages which are possessed by a mixture of peace and terror.

"Oh, yes, we shot seventeen of them."

They had shot seventeen "fascists." The parish priest, the priest's housekeeper, the sexton, and fourteen village notables. Everything is relative, you see. When they read in their provincial newspaper the story of the life of Basil Zaharoff, master of the world, they transpose it into their own language. They recognize in him the nurseryman, or the pharmacist. And when they shoot the pharmacist, in a way they are shooting Basil Zaharoff. The only one who does not understand is the pharmacist.

"Now we are all Loyalists together. Everything has calmed down."

Almost everything. The conscience of the village is tormented by one man whom I have seen at the tavern, smiling, helpful, so anxious to go on living! He comes to the pub in order to show us that, despite his few acres of vineyard, he too is part of the human race, suffers with rheumatism like it, mops his face like it with a blue handkerchief. He comes, and he plays billiards. Can one shoot a man who plays billiards? Besides, he plays badly with his great trembling hands. He is upset; he still does not know whether he is a fascist or not. He puts me in mind of those poor monkeys who dance before the boa-constrictor in the hope of softening it.

There was nothing we could do for the man. For the time being we had another job in hand. Sitting on a table and swinging my legs at committee headquarters, while my companion, Pépin, pulled a bundle of soiled papers out of his pocket, I had a good look at these terrorists. Their looks belied their name: honorable peasants with frank eyes and sober attentive faces, they were the same everywhere we went; and though we were foreigners possessing no authority, we were everywhere received with the same grave courtesy.

"Yes, here it is," said Pépin, a document in his hand. "His name is Laporte. Any of you know him?"

The paper went from hand to hand and the members of the committee shook their heads.

"No. Laporte? Never heard of him."

I started to explain something to them, but Pépin motioned me to be silent. "They won't talk," he said, "but they know him well enough."

Pépin spread his references before the chairman, saying casually: "I am a French socialist. Here is my party card."

The card was passed round and the chairman raised his eyes to us:

"Laporte. I don't believe. . . ."

"Of course you know him. A French monk. Probably in disguise. You captured him yesterday in the woods. Laporte, his name is. The French consulate wants him."

I sat swinging my legs. What a strange session! Here we were in a mountain village sixty miles from the French frontier, asking a revolutionary committee that shot even parish priests' housekeepers to surrender to us in good shape a French monk. Whatever happened to us, we would certainly have asked for it. Nevertheless, I felt safe. There was no treachery in these people. And why, as a matter of fact, should they bother to play tricks? We had absolutely no protection; we meant no more to them than Laporte; they could do anything they pleased.

Pépin nudged me. "I've an idea we have come too late," he said.

The chairman cleared his throat and made up his mind.

"This morning," he said, "we found a dead man on the road just outside the village. He must be there still."

And he pretended to send off for the dead man's papers.

"They've already shot him," Pépin said to me. "Too bad! They would certainly have turned him over to us. They are good kind people."

I looked straight into the eyes of these curious "good kind people." Strange: there was nothing in their eyes to upset me. There seemed nothing to fear in their set jaws and the blank smoothness of their faces. Blank, as if vaguely bored. A rather terrible blankness. I wondered why, despite our unusual mission, we were not suspect to them. What difference had they established in their minds between us and the "fascist" in the neighboring tavern who was dancing his dance of death before the unavailing indifference of these judges? A crazy notion came into my head, forced upon my attention by all the power of my instinct: If one of those men yawned I should be afraid. I should feel that all human communication had snapped between us.

After we left, I said to Pépin:

"That is the third village in which we have done this job and I still cannot make up my mind whether the job is dangerous or not."

Pépin laughed and admitted that although he had saved dozens of men on these missions, he himself did not know the answer.

"Yesterday," he confessed, "I had a narrow squeak. I snaffled a Carthusian monk away from them just as they were about to shoot the fellow. The smell of blood was in the air, and . . . Well, they growled a bit, you know."

I know the end of that story. Pépin, the socialist and notorious anti-church political worker, having staked his life to get that Carthusian, had hustled him into a motor-car and there, by way of compensation, he sought to insult the priest by the finest bit of blasphemy he could summon:

"You . . . you . . . you triple damned monk!" he had finally spluttered.

This was Pépin's triumph. But the monk, who had not been listening, flung his arms round Pépin's neck and wept with happiness.

In another village they gave up a man to us. With a great air of mystery, four militiamen dug him up out of a cellar. He was a lively bright-eyed monk whose name I have already forgotten, disguised as a peasant and carrying a long gnarled stick scarred with notches.

"I kept track of the days," he explained. "Three weeks in the woods is a long time. Mushrooms are not specially nourishing, and they grabbed me when I came near a village.

The mayor of the village, to whom we owed this gift, was very proud of him.

"We shot at him a lot and thought we had killed him," he said.

And then, by way of excuse for the bad marksmanship, he added: "I must say it was at night."

The monk laughed.

"I wasn't afraid."

We put him into the car, and before we threw in the clutch everybody had to shake hands all round with these terrible terrorists. The monk's hand was shaken hardest of all and he was repeatedly congratulated on being alive. To all these friendly sentiments he responded with a warmth of unquestionably sincere appreciation.

As for me, I wish I understood mankind.

We went over our lists. At Sitges lived a man who, we had been told, was in danger of being shot. We drove round and found his door wide open. Up a flight of stairs we ran into our skinny young man.

"It seems that these people are likely to shoot you," we told him. "Come back to Barcelona with us and you will be shipped home to France in the *Duquesne*."

The young man took a long time to think this over and then said:

"This is some trick of my sister's."

"What?"

"She lives in Barcelona. She would never pay for the child's keep and I always had to. . . ."

"Your family troubles are none of our affair. Are you in danger here, yes or no?"

"I don't know. I tell you, my sister . . ."

"Do you want to get away, yes or no?"

"I really don't know. What do you think? In Barcelona, my sister . . ."

The man was carrying on his family quarrel through the revolution. He was going to stay here in order to do his sister in the eye.

"Do as you please," we said, finally, and we left him where he was.

We stopped the car and got out. A volley of rifle-shot had crackled in the still country air. From the top of the road we looked down upon a clump of trees out of which, a quarter of a mile away, stuck two tall chimneys. A squad of militiamen came up and loaded their guns. We asked what was going on. They looked round, pointed to the chimneys, and decided that the firing must have come from the factory.

The shooting died down almost immediately, and silence fell again. The chimneys went on smoking peacefully. A ripple of wind ran over the grass. Nothing had changed visibly, and we ourselves were unchanged. Nevertheless, in that clump of trees someone had just died.

One of the militiamen said that a girl had been killed at the factory, together with her brothers, but there was still some uncertainty about this. What excruciating simplicity! Our own peace of mind had not been invaded by those muffled sounds in the clump of greenery, by that brief partridge drive. The angelus, as it were, that had rung out in that foliage had left us calm and unrepentant.

Human events display two faces, one of drama and the other of indifference. Everything changes according as the event concerns the individual or the species. In its migrations, in its imperious impulses, the species forgets its dead. This, perhaps, explains the unperturbed faces of these peasants. One feels that they have no special taste for horror; yet they will come back from that clump of trees on the one hand content to have administered their kind of justice, and on the other hand quite indifferent to the fact of the girl who stumbled against the root of the tree of death, who was caught by death's harpoon as she fled, and who now lies in the wood, her mouth filled with blood.

Here I touch the inescapable contradiction I shall never be able to resolve. For man's greatness does not reside merely in the destiny of the species: each individual is an empire. When a mine caves in and closes over the head of a single miner, the life of the community is suspended.

His comrades, their women, their children, gather in anguish at the entrance to the mine, while below them the rescue party scratch with their picks at the bowels of the earth. What are they after? Are they consciously saving one unit of society? Are they freeing a human being as one might free a horse, after computing the work he is still capable of doing? Ten other miners may be killed in the attempted rescue: what inept cost accounting! Of course it is not a matter of saving one ant out of the colony of ants! They are rescuing a consciousness, an empire whose significance is incommensurable with anything else.

Inside the narrow skull of the miner pinned beneath the fallen timber, there lives a world. Parents, friends, a home, the hot soup of evening, songs sung on feast days, loving kindness and anger, perhaps even a social consciousness and a great universal love, inhabit that skull. By what are we to measure the value of a man? His ancestor once drew a reindeer on the wall of a cave; and two hundred thousand years later that gesture still radiates. It stirs us, prolongs itself in us. Man's gestures are an eternal spring. Though we die for it, we shall bring up that miner from his shaft. Solitary he may be; universal he surely is.

In Spain there are crowds in movement, but the individual, that universe, calls in vain for help from the bottom of the mine.

## III

Machine-gun bullets cracked against the stone above our heads as we skirted the moonlit wall. Low-flying lead thudded into the rubble of an embankment that rose on the other side of the road. Half a mile away a battle was in progress, the line of fire drawn in the shape of a horse-shoe ahead of us and on our flanks.

Walking between wall and parapet on the white highway, my guide and I were able to disregard the spatter of missiles in a feeling of perfect security. We could sing, we could laugh, we could strike matches, without drawing upon ourselves the direct fire of the enemy. We went forward like peasants on their way to market. Half a mile away the iron hand of war would have set us inescapably upon the black chessboard of battle; but here, out of the game, ignored, the Republican lieutenant and I were as free as air.

Shells filled the night with absurd parabolas during their three seconds of freedom between release and exhaustion. There were the duds that dove without bursting into the ground; there were the travelers in space that whipped straight overhead, elongated in their race to the stars. And the leaden bullets that ricocheted in our faces and tinkled curiously in our ears were like bees, dangerous for the twinkling of an eye, poisonous but ephemeral.

Walking on, we reached a point where the embankment had collapsed.

"We might follow the cross-trench from here," my guide suggested.

Things had suddenly turned serious. Not that we were in the line of machine-gun fire, or that a roving searchlight was about to spot us. It was not as bad as that. There had simply been a rustling overhead; a sort of celestial gurgle had sounded. It meant no harm to us, but the lieutenant remarked suddenly, "That is meant for Madrid," and we went down into the trench.

The trench ran along the crest of a hill a little before reaching the suburb of Carabanchel. In the direction of Madrid a part of the parapet had crumbled and we could see the city in the gap, white, strangely white, under the full moon. Hardly a mile separated us from those tall structures dominated by the tower of the Telephone Building.

Madrid was asleep—or rather Madrid was feigning sleep. Not a light; not a sound. Like clockwork, every two minutes the funereal fracas that we were henceforth to hear roared forth and was dissolved in a dead silence. It seemed to waken no sound and no stirring in the city, but was swallowed up each time like a stone in water.

Suddenly in the place of Madrid I felt that I was staring at a face with closed eyes. The hard face of an obstinate virgin taking blow after blow without a moan. Once again there sounded overhead that gurgling in the stars of a newly uncorked bottle. One second, two seconds, five seconds went by. There was an explosion and I ducked involuntarily. There goes the whole town, I thought.

But Madrid was still there. Nothing had collapsed. Not an eye had blinked. Nothing was changed. The stone face was as pure as ever.

"Meant for Madrid," the lieutenant repeated mechanically. He taught me to tell these celestial shudders apart, to follow the course of these sharks rushing upon their prey:

"No, that is one of our batteries replying. . . . That's theirs, but firing somewhere else. . . . There's one meant for Madrid."

Waiting for an explosion is the longest passage of time I know.

What things go on in that interminable moment! An enormous pressure rises, rises. Will that boiler ever make up its mind to burst? At last! For some that meant death, but there are others for whom it meant escape from death. Eight hundred thousand souls, less half a score of dead, have won a last-minute reprieve. Between the gurgling and the explosion eight hundred thousand lives were in danger of death.

Each shell in the air threatened them all. I could feel the city out there, tense, compact, a solid. I saw them all in the mind's eye —men, women, children, all that humble population crouching in the sheltering cloak of stone of a motionless virgin. Again I heard the ignoble crash and was gripped and sickened by the downward course of the torpedo. . . . Torpedo? I scarcely knew what I was saying. "They . . . they are torpedoing Madrid." And the lieutenant, standing there counting the shells, said:

"Meant for Madrid. Sixteen."

I crept out of the trench, lay flat on my stomach on the parapet, and stared. A new image has wiped out the old. Madrid with its chimney-pots, its towers, its portholes, now looks like a ship on the high seas. Madrid all white on the black waters of the night. A city outlives its inhabitants. Madrid, loaded with emigrants, is ferrying them from one shore to the other of life. It has a generation on board. Slowly it navigates through the centuries. Men, women, children fill it from garret to hold. Resigned or quaking with fear, they live only for the moment to come. A vessel loaded with humanity is being torpedoed. The purpose of the enemy is to sink Madrid as if she were a ship.

Stretched out on the parapet I do not care a curse for the rules of war. For justifications or for motives. I listen. I have learned to read the course of these gurglings among the stars. They pass quite close to Sagittarius. I have learned to count slowly up to five.

And I listen. But what tree has been sundered by this lightning, what cathedral has been gutted, what poor child has just been stricken, I have no means of knowing.

That same afternoon I had witnessed a bombardment in the town itself. All the force of this thunder-clap had to burst on the Gran Via in order to uproot a human life. One single life. Passers-by had brushed rubbish off their clothes; other had scattered on the run; and when the light smoke had risen and cleared away, the betrothed, escaped by miracle without a scratch, found at his feet his *novia,* whose golden arm a moment before had been in his, changed into a blood-filled sponge, changed into a limp packet of flesh and rags.

He had knelt down, still uncomprehending, had nodded his head slowly, as if saying to himself, "Something very strange has happened."

This marvel spattered on the pavement bore no resemblance to what had been his beloved. Misery was excruciatingly slow to engulf him in its tidal wave. For still another second, stunned by the feat of the invisible prestidigitator, he cast a bewildered glance round him in search of the slender form, as if it at least should have survived. Nothing was there but a packet of muck.

Gone was the feeble spark of humanity. And while in the man's throat there was brewing that shriek which I know not what deferred, he had the leisure to reflect that it was not those lips he had loved but their pout, not them but their smile. Not those eyes, but their glance. Not that breast, but its gentle swell. He was free to discover at last the source of the anguish love had been storing up for him, to learn that it was the unattainable he had been pursuing. What he had yearned to embrace was not the flesh but a downy spirit, a spark, the impalpable angel that inhabits the flesh.

I do not care a curse for the rules of war and the law of reprisal.

As for the military advantage of such a bombardment, I simply cannot grasp it. I have seen housewives disemboweled, children mutilated; I have seen the old itinerant market crone sponge from her treasures the brains with which they were spattered. I have seen a janitor's wife come out of her cellar and douse the sullied pavement with a bucket of water, and I am still unable to understand what part these humble slaughter-house accidents play in warfare.

A moral rôle? But a bombardment turns against the bombarder! Each shell that fell upon Madrid fortified something in the town. It persuaded the hesitant neutral to plump for the defenders. A dead child weighs heavily in the balance when it is one's own. It was clear to me that a bombardment did not disperse—it unified. Horror causes men to clench their fists, and in horror men join together.

The lieutenant and I crawled along the parapet. Face or ship, Madrid stood erect, receiving blows without a moan. But men are like this: slowly but surely, ordeal fortifies their virtues.

Because of the ordeal my companion's heart was high. He was thinking of the hardening of Madrid's will. He stood up with his fists on his hips, breathing heavily. Pity for the women and the children had gone out of him.

"That makes sixty," he counted grimly.

The blow resounded on the anvil. A giant smith was forging Madrid.

One side or the other would win. Madrid would resist or it would fall. A thousand forces were engaged in this mortal confusion of tongues from which anything might come forth. But one did not need to be a Martian, did not need to see these men dispassionately in a long perspective, in order to perceive that they were struggling against themselves, were their own enemy. Mankind perhaps was being brought to bed of something here in Spain; something perhaps was to be born of this chaos, this disruption. For indeed not all that I saw in Spain was horror, not all of it filled my mouth with a taste of ashes.

## IV

On the Guadalajara front I sat at night in a dugout with a Republican squad made up of a lieutenant, a sergeant, and three men. They were about to go out on patrol duty. One of them—the night was cold—stood half in shadow with his head not quite through the neck of a sweater he was pulling on, his arms caught in the sleeves and waving slowly and awkwardly in the air like the short arms of a bear. Smothered curses, stubbles of beard, distant muffled explosions—the atmosphere was a strange compound of sleep, waking, and death. I thought of tramps on the road bestirring themselves, raising themselves up off the ground on heavy sticks. Caught in the earth, painted by the earth, their hands grubby with their gardenless gardening, these men were raising themselves painfully out of the mud in order to emerge under the stars. In these blocks of caked clay I could sense the awakening of consciousness, and as I looked at them I said to myself that across the way, at this very moment, the enemy was getting into his harness, was thickening his body with woolen sweaters; earth-crusted, he was breaking out of his mould of hardened mud. Across the way the same clay shaping the same beings was wakening in the same way into consciousness.

The patrol moved forward across fields through crackling stubble, knocking its toes against unseen rocks in the dark. We were making our way down into a narrow valley on the other side of which the enemy was entrenched. Caught in the cross-fire of artillery, the peasants had evacuated this valley, and their deserted village lay here drowned in the waters of war. Only their dogs remained, ghostly creatures that hunted their pitiful prey in the day and howled in the night. At four in the morning, when the moon rose white as a picked bone, a whole village bayed at the dead divinity.

"Go down and find out if the enemy is hiding in that village," the commanding officer had ordered. Very likely on the other side the same order had been given.

We were accompanied by a sort of political agent, a civilian, whose name I have forgotten, though not what he looked like. It seems to me he must have been rheumatic, and I remember that he leaned heavily on a knotted stick as we tramped forward in the night. His face was the face of a conscientious and elderly workman. I would have sworn that he was above politics and parties, above ideological rivalries. "Pity it is," he would say, "that as things are we cannot explain our point of view to the other fellow." He walked weighed down by his doctrine, like an evangelist. Across the way, meanwhile, was the other evangelist, a believer just as enlightened as this one, his boots just as muddy, his duty taking him on exactly the same errand.

"You'll hear them pretty soon," my commissar said. "When we get close enough we'll call out to the enemy, ask him questions; and he may answer tonight."

Although we don't yet know it, we are in search of a gospel to embrace all gospels, we are on the march towards a stormy Sinai.

And we have arrived. Here is a dazed sentry, half asleep in the window of a stone wall.

"Yes," says my commissar, "sometimes they answer. Sometimes they call out first and ask questions. Of course they don't answer, too, sometimes. Depends on the mood they're in."

Just like the gods.

A hundred yards behind us lie our trenches. I strike a match, intending to light a cigarette, and two powerful hands duck my head. Everybody has ducked, and I hear the whistle of bullets in the air. Then silence. The shots were fired high and the volley was not repeated—a mere reminder from the enemy of what constitutes decorum here. One does not light a cigarette in the face of the enemy.

We are joined by three or four men, wrapped in blankets, who had been posted behind neighboring walls.

"Looks as if the lads across the way were awake," one of them remarks.

"Do you think they'll talk tonight? We'd like to talk to them."

"One of them, Antonio, he talks sometimes."

"Call him."

The man in the blanket straightens up, cups his hands round his mouth, takes a deep breath, and calls out slowly and loudly: "An . . . to . . . ni . . . o!"

The call swells, unfurls, floats across the valley and echoes back.

"Better duck," my neighbor advises. "Sometimes when you call them, they let fly."

Crouched behind the stone wall, we listen. No sound of a shot. Yet we cannot say we have heard nothing at all, for the whole night is singing like a sea-shell.

"Hi! Antonio . . . o! Are you . . . "

The man in the blanket draws another deep breath and goes on: "Are you asleep?"

"Asleep?" says the echo. "Asleep?" the valley asks. "Asleep?" the whole night wants to know. The sound fills all space. We scramble to our feet and stand erect in perfect confidence. They have not touched their guns.

I stand imagining them on their side of the valley as they listen, hear, receive this human voice, this voice that obviously has not stirred them to anger since no finger has pressed a trigger. True, they do not answer, they are silent; but how attentive must be that silent audience from which, a moment ago, a match had sufficed to draw a volley. Borne on the breeze of a human voice, invisible seeds are fertilizing that black earth across the valley. Those men thirst for our words as we for theirs. But their fingers, meanwhile, are on their triggers. They put me in mind of those wild things we would try in the desert to tame and that would stare at us, eat the food and drink the water we set out for them, and would spring at our throats when we made a move to stroke them.

We squatted well down behind the wall and held up a lighted match above it. Three bullets passed overhead. To the match they said, "You are forgetting that we are at war." To us, "We are listening, nevertheless. We can still love, though we stick to our rules."

Another giant peasant rested his gun against the wall, stood up, drew a deep breath, and let go:

"Antonio . . . o! It's me! Leo!"

The sound moved across the valley like a ship new-launched. Eight hundred yards to the far shore, eight hundred back—sixteen hundred yards. If they answered, there would be five seconds of time between our questions and their replies. Five seconds of silence, in which all war would be suspended, would go by between each question and each answer. Like an embassy on a journey, each time. What this meant was that even if they answered, we should still feel ourselves separated from them. Between them and us the inertia of an invisible world would still be there to be stirred into action. For the considerable space of five seconds we should be like men shipwrecked and fearful lest the rescue party had not heard their cries.

". . . ooo!"

A distant voice like a feeble wave has curled up to die on our shore. The phrase, the word, was lost on the way and the result is an undecipherable message. Yet it strikes me like a blow. In this impenetrable darkness a sudden flash of light has gleamed. All of us are shaken by a ridiculous hope. Something has made known to us its existence. We can be sure now that there are men across the way. It is as if in invisibility a crack had opened, as if . . . Imagine a house at night, dark and its doors all locked. You, sitting in its darkness, suddenly feel a breath of cold air on your face. A single breath. What a presence!

There it comes again! ". . . time . . . sleep!"

Torn, mutilated as a truly urgent message must be, washed by the waves and soaked in brine, here is our message. The men fired at our cigarettes have blown up their chests with air in order to send us this mother bit of advice:

"Quiet! Go to bed! Time to sleep!"

It excites us. You who read this will perhaps think that these men were merely playing a game. In a sense they were. I am sure that, being simple men, if you had caught them at their sport they would have denied that it was serious. But games always cover something deep and intense, else there would be no excitement in them, no pleasure, no power to stir us. Here was a game that made our hearts beat too wildly not to satisfy a real though undefined need within us. It was as if we were marrying our enemy before dying of his blow.

But so slight, so fragile was the pontoon flung between our two shores that a question too awkward, a phrase too clumsy, would certainly upset it. Words lose themselves: only essential words, only the truth of truths would leave this frail bridge whole. And I can see him now, that peasant who stirred Antonio to speech and thus made himself our pilot, our ambassador; I can see him as he stood erect, as he rested his strong hands on the low stone wall and sent forth from his great chest that question of questions:

"Antonio! What are you fighting for?"

Let me say again that he and Antonio would be ashamed to think that you took them seriously. They would insist that it was all in fun. But I was there as he stood waiting, and I know that his whole soul gaped wide to receive the answer. Here is the truncated message, the secret mutilated by five seconds of travel across the valley as an inscription in stone is defaced by the passing of the centuries:

". . . Spain!"

And then I heard:

". . . You?"

He got his answer. I heard the great reply as it was flung forth into space:

"The bread of our brothers!"

And then the amazing:

"Good night, friend!"

And the response from the other side of the world:

"Good night, friend!"

And silence.

Their words were not the same, but their truths were identical. Why has this high communion never yet prevented men from dying in battle against each other?

## V

Back on the Madrid front I sat again at night in a subterranean chamber, at supper with a young captain and a few of his men. The telephone had rung and the captain was being ordered to prepare to attack before daybreak. Twenty houses in this industrial suburb, Carabanchel, constituted the objective. There would be no support: one after the other the houses were to be blown in with hand grenades and occupied.

I felt vaguely squeamish as I took something like a last look at these men who were shortly to dive into the great bowl of air, suck the blue night into their lungs, and then be blown to bits before they could reach the other side of the road. They were taking it easily enough, but the captain came back to table from the telephone shrugging his shoulders. "The first man out . . ." He started to say something, changed his mind, pushed two glasses and a bottle of brandy across the table, and said to the sergeant:

"You lead the file with me. Have a drink and go get some sleep."

The sergeant drank and went off to sleep. Round the table a dozen of us were sitting up. All the chinks in this room were caulked up; not a trickle of light could escape; the glare within was so dazzling that I blinked. The brandy was sweet, faintly nauseating, and its taste was as mournful as a drizzle at daybreak. I was in a daze, and when I had drunk I shut my eyes and saw behind my lids those ruined and ghostly houses bathed in a greenish radiance as of moonglow under water, that I had stared at a few minutes before through the sentry's loophole. Someone on my right was telling a funny story. He was talking very fast and I understood about one word in three.

A man came in half drunk, reeling gently in this half-real world. He stood rubbing a stubble of beard and looking us over with vague affectionate eyes. His glance slid across to the bottle, avoided it, came back to it, and turned pleadingly to the captain.

The captain laughed softly, and the man, suddenly hopeful, laughed too. A light gust of laughter ran over the roomful of men.

The captain put out his hand and moved the bottle noiselessly out of reach. The man's glance simulated despair, and a childish game began, a sort of mute ballet which, in the fog of cigarette smoke and the weariness of the watch with its anticipation of the coming attack, was utterly dream-like. I sat hypnotized by this atmosphere of the slowly ending vigil, reading the hour in the stubbles of beard while out of doors a sea-like pounding of cannon waxed in intensity.

Soon afterwards these men were to scour themselves clean of their sweat, their brandy, the filth of their vigil, in the regal waters of the night of war. I felt in them something so near to spotless purity! Meanwhile, as long as it would last, they were dancing the ballet of the drunkard and the bottle. They were determined that this game should absorb them utterly. They were making life last as long as it possibly could. But there on a shelf stood a battered alarm clock, set to sound the zero-hour. No one so much as glanced at it but me, and my glance was furtive. They would all hear it well enough, never fear! Its ringing would shatter the stifling air.

The clock would ring out. The men would rise to their feet and stretch themselves. They would be sure to make this gesture which is instinctive in every man about to tackle the problem of survival. They would stretch themselves, I say, and they would buckle on their harness. The captain would pull his revolver out of his holster. The drunk would sober up. And all these men, without undue haste, would file into the passage. They would go as far as that rectangle of pale light which is the sky at the end of the passage, and there they would mutter something simple like "Look at that moon!" or "What a night!" And then they would fling themselves into the stars.

Scarcely had the attack been called off by telephone, scarcely had these men, most of whom had been doomed to die in the attack upon that concrete wall, begun to feel themselves safe, begun to realize that they were certain of trampling their sweet planet in their rough clogs one more day, scarcely were their minds at peace, when all in chorus began to lament their fate.

"Do they think we are a lot of women?" "Is this a war or isn't it?" "A fine general staff!" they grumbled sarcastically. Can't make up its mind about anything! Wants to see Madrid bombarded and kids smashed to bits. Here they were, ready to rip up those enemy batteries and fling them over the backs of mountains to save innocence imperiled, and the staff tied them hand and foot, condemned them to inaction.

It was clear enough, and the men admitted it, that none of them might have come up again after their dive into the moonlight, and that they ought in reality to be very happy to be alive and able to grouse against G. H. Q. and go on drinking their consoling brandy;—and, by the way, since the second telephone message, two curious things had happened: the brandy tasted better and the men were now drinking it cheerfully instead of moodily.

Yet at the same time I saw nothing in their vehemence that made me think it either silly or boastful. I could not but remember that all of them had been ready to die with simplicity.

Day broke. I scrubbed my face in the freezing water of the village pump. Coffee steamed in the bowls under an arbor forty yards from the enemy outpost, half-wrecked by the midnight firing but safe in the truce of dawn. Now freshly washed, the survivors gathered here to commune in life rather than in death, to share their white bread, their cigarettes, their smiles. They came in one by one, the captain, Sergeant R——, the lieutenant, and the rest, planted their elbows solidly on the table, and sat facing this treasure which they had been judicious enough to despise at a moment when it seemed it must be abandoned, but which had now recovered its price. *"Salud, amigo!"*—"Hail, friend!"—they sang out as they clapped one another on the shoulder.

I loved the freezing wind that caressed us and the shining sun that warmed us beneath the touch of the wind. I loved the mountain air that was filling me with gladness. I rejoiced in the cheer of these men who sat in their shirtsleeves gathering fresh strength from their repast and making ready, once they had finished and risen to their feet, to knead the stuff of the world.

A ripe pod burst somewhere. From time to time a silly bullet spat against the stone wall. Death was abroad, of course, but wandering aimlessely and without ill intent. This was not death's hour. We in the arbor were celebrating life.

This whole platoon had risen up *de profundis;* and the captain sat breaking the white bread, that densely baked bread of Spain so rich in wheat, in order that each of his comrades, having stretched forth his hand, might receive a chunk as big as his fist and turn it into life.

These men had in truth risen *de profundis*. They were in very fact beginning a new life. I stared at them, and in particular at Sergeant R——, he who was to have been the first man out and who had gone to sleep in preparation for the attack. I was with them when they woke him up. Now Sergeant R—— had been well aware that he was to be the first man to step out into the line of fire of a machine-gun nest and dance in the moonlight that brief ballet at the end of which is death. His awakening had been the awakening of a prisoner in the death cell.

At Carabanchel the trenches wound among little workmen's houses whose furnishings were still in place. In one of these, a few yards from the enemy, Sergeant R—— was sleeping fully dressed on an iron cot. When we had lighted a candle and had stuck it into the neck of a bottle, and had drawn forth out of the darkness that funereal bed, the first thing that came into view was a pair of clogs. Enormous clogs, iron-shod and studded with nails, the clogs of a sewer-worker or a railway trackwalker. All the poverty of the world was in those clogs. No man ever strode with happy steps through life in clogs like these: he boarded life like a longshoreman for whom life is a ship to be unloaded.

This man was shod in his tools, and his whole body was covered with the tools of his trade—cartridge belt, gun, leather harness. His neck was bent beneath the heavy collar of the draught horse. Deep in caves, in Morocco, you can see millstones worked by blind horses. Here in the ruddy wavering light of the candle we were waking up a blind horse and sending him out to the mill.

"Hi! Sergeant!"

He sent forth a sigh as heavy as a wave and turned slowly and massively over towards us so that we saw a face still asleep and filled with anguish. His eyes were shut, and his mouth, to which clung a bubble of air, was half open like the mouth of a drowned man. We sat down on his bed and watched his laborious awakening. The man was clinging like a crab to submarine depths, grasping in his fists I know not what dark seaweed. He opened and shut his hands, pulled up another deep sigh, and escaped from us suddenly with his face to the wall, obstinate with the stubbornness of an animal refusing to die, turning its back on the slaughterhouse.

"Hi! Sergeant!"

Once again he was drawn up from the bottom of the sea, swam towards us, and we saw again his face in the candle-light. This time we had hobbled our sleeper; he would not get away from us again. He blinked with closed eyes, moved his mouth round as if swallowing, ran his hand over his forehead, made one great effort to sink back into his happy dreams and reject our universe of dynamite, weariness, and glacial night, but it was too late. Something from without was too strong for him.

Like the punished schoolboy stirred by the insistent bell out of his dream of a school-less world, Sergeant R—— began to clothe himself in the weary flesh he had so recently shed, that flesh which in the chill of awakening was soon to know the old pains in the joints, the weight of the harness, and the stumbling race towards death. Not so much death as the discomfort of dying, the filth of the blood in which he would steep his hands when he tried to rise to his feet; the stickiness of that coagulating syrup. Not so much death as the Calvary of a punished child.

One by one he stretched his arms and then his legs, bringing up an elbow, straightening a knee, while his straps, his gun, his cartridge belt, the three grenades hanging from his belt, all hampered the final strokes of this swimmer in the sea of sleep. At last he opened his eyes, sat up on the bed, and stared at us, mumbling:

"Huh! Oh! Are we off?"

And as he spoke, he simply stretched out his hand for his rifle.

"No," said the captain. "The attack has been called off."

Sergeant R——, let me tell you that we made you a present of your life. Just that. As much as if you had stood at the foot of the electric chair. And God knows, the world sheds ink enough on the pathos of pardon at the foot of the electric chair. We brought you your pardon *in extremis*. No question about it. In your mind there was nothing between you and death but a thickness of tissue-paper. Therefore you must forgive me my curiosity. I stared at you, and I shall never forget your face. It was a face touching and ugly, with a humped nose a little too big, high cheek-bones, and the spectacles of an intellectual. How does a man receive the gift of life? I can answer that. A man sits still, pulls a bit of tobacco out of his pocket, nods his head slowly, looks up at the ceiling, and says:

"Suits me."

Then he nods his head again and adds:

"If they'd sent us a couple of platoons the attack might have made sense. The lads would have pitched in. You'd have seen what they can do."

Sergeant, Sergeant, what will you do with this gift of life?

Now, Sergeant at peace, you are dipping your bread into your coffee. You are rolling cigarettes. You are like the lad who has been told he will not be punished after all. And yet, like the rest, you are ready to start out again tonight on that brief dash at the end of which the only thing a man can do is kneel down.

Over and over in my head there goes the question I have wanted to ask you ever since last night: "Sergeant, what is it makes you willing to die?"

But I know that it is impossible to ask such a question. It would offend a modesty in you which you yourself do not know to be there, but which would never forgive me. You could not answer with high-sounding words: they would seem false to you and in truth they would be false. What language could be chaste enough for a modest man like you? But I am determined to know, and I shall try to get round the difficulty. I shall ask you seemingly idle questions, and you will answer.

"Tell me, why did you join up?"

If I understood your answer, Sergeant, you hardly know yourself. You were a bookkeeper in Barcelona. You added up your columns of figures every day without worrying much about the struggle against the rebels. But one of your friends joined up, and then a second friend; and you were disturbed to find yourself undergoing a curious transformation: little by little your columns of figures seemed to you futile. Your pleasures, your work, your dreams, all seemed to belong to another age.

But even that was not important, until one day you heard that one of your friends had been killed on the Málaga front. He was not a friend for whom you would ever have felt you had to lay down your life. Yet that bit of news swept over you, over your narrow little life, like a wind from the sea. And that morning another friend had looked at you and said, "Do we or don't we?" And you had said, "We do."

You never really wondered about the imperious call that compelled you to join up. You accepted a truth which you could never translate into words, but whose self-evidence overpowered you. And while I sat listening to your story, an image came into my mind, and I understood.

When the wild ducks or the wild geese migrate in their season, a strange tide rises in the territories over which they sweep. As if magnetized by the great triangular flight, the barnyard fowl leap a foot or two into the air and try to fly. The call of the wild strikes them with the force of a harpoon and a vestige of savagery quickens their blood. All the ducks on the farm are transformed for an instand into migrant birds, and into those hard little heads, till now filled with humble images of pools and worms and barnyards, there swims a sense of continental expanse, of the breadth of seas and the salt taste of the ocean wind. The duck totters to right and left in its wire enclosure, gripped by a sudden passion to perform the impossible and a sudden love whose object is a mystery.

Even so is man overwhelmed by a mysterious presentiment of truth, so that he discovers the vanity of his bookkeeping and the emptiness of his domestic felicities. But he can never put a name to this sovereign truth. Men explain these brusque vocations by the need to escape or the lure of danger, as if we knew where the need to escape and the lure of danger themselves came from. They talk about the call of duty, but what is it that makes the call of duty so pressing? What can you tell me, Sergeant, about that uneasiness that seeped in to disturb your peaceful existence?

The call that stirred you must torment all men. Whether we dub it sacrifice, or poetry, or adventure, it is always the same voice that calls. But domestic security has succeeded in crushing out that part in us that is capable of heeding the call. We scarcely quiver; we beat our wings once or twice and fall back into our barnyard.

We are prudent people. We are afraid to let go of our petty reality in order to grasp at a great shadow. But you, Sergeant, did discover the sordidness of those shopkeepers' bustlings, those petty pleasures, those petty needs. You felt that men did not live like this. And you agreed to heed the great call without bothering to try to understand it. The hour had come when you must moult, when you must rise into the sky.

The barnyard duck had no notion that his little head was big enough to contain oceans, continents, skies; but of a sudden here he was beating his wings, despising corn, despising worms, battling to become a wild duck.

There is a day of the year when the eels must go down to the Sargasso Sea, and come what may, no one can prevent them. On that day they spit upon their ease, their tranquillity, their tepid waters. Off they go over ploughed fields, pricked by the hedges and skinned by the stones, in search of the river that leads to the abyss.

Even so did you feel yourself swept away by that inward migration about which no one had ever said a word to you. You were ready for a sort of bridal that was a mystery to you, but in which you had to participate. "Do we or don't we? We do." You went up to the front in a war that at bottom meant little to you. You took to the road as spontaneously as that silvery people shining in the fields on its way to the sea, or that black triangle in the sky.

What were you after? Last night you almost reached your goal. What was it you discovered in yourself that was so ready to burst from its cocoon? At daybreak your comrades were full of complaint: tell me, of what had they been defrauded? What had they discovered in themselves that was about to show itself, and that now they wept for?

What, Sergeant, were the visions that governed your destiny and justified your risking your life in this adventure? Your life, your only treasure! We have to live a long time before we become men. Very slowly do we plait the braid of friendships and affections. We learn slowly. We compose our creation slowly. And if we die too early we are in a sense cheated out of our share. We have to live a long time to fulfil ourselves.

But you, by the grace of an ordeal in the night which stripped you of all that was not intrinsic, you discovered a mysterious creature born of yourself. Great was this creature, and never shall you forget him. And he is yourself. You have had the sudden sense of fulfilling yourself in the instant of discovery, and you have learned suddenly that the future is now less necessary for the accumulation of treasures. That creature within you who opened his wings is not bound by ties to perishable things; he agrees to die for all men, to be swallowed up in something universal.

A great wind swept through you and delivered from the matrix the sleeping prince you sheltered—Man within you. You are the equal of the musician composing his music, of the physicist extending the frontier of knowledge, of all those who build the highways over which we march to deliverance. Now you are free to gamble with death. What have you now to lose?

Let us say you were happy in Barcelona: nothing more can ruin that happiness. You have reached an altitude where all loves are of the same stuff. Perhaps you suffered on earth, felt yourself alone on the planet, knew no refuge to which you might fly? What of that! Sergeant, this day you have been welcomed home by love.

## VI

No man can draw a free breath who does not share with other men a common and disinterested ideal. Life has taught us that love does not consist in gazing at each other but in looking outward together in the same direction. There is no comradeship except through union in the same high effort. Even in our age of material well-being this must be so, else how should we explain the happiness we feel in sharing our last crust with others in the desert? No sociologist's textbook can prevail against this fact. Every pilot who has flown to the rescue of a comrade in distress knows that all joys are vain in comparison with this one. And this, it may be, is the reason why the world today is tumbling about our ears. It is precisely because this sort of fulfilment is promised each of us by his religion, that men are inflamed today. All of us, in words that contradict each other, express at bottom the same exalted impulse. What sets us against one another is not our aims—they all come to the same thing—but our methods, which are the fruit of our varied reasoning.

Let us, then, refrain from astonishment at what men do. One man finds that his essential manhood comes alive at the sight of self-sacrifice, cooperative effort, a rigorous vision of justice, manifested in an anarchists' cellar in Barcelona. For that man there will henceforth be but one truth—the truth of the anarchists. Another, having once mounted guard over a flock of terrified little nuns kneeling in a Spanish nunnery, will thereafter know a different truth—that it is sweet to die for the Church. If, when Mermoz plunged into the Chilean Andes with victory in his heart, you had protested to him that no merchant's letter could possibly be worth risking one's life for, Mermoz would have laughed in your face. Truth is the man that was born in Mermoz when he slipped through the Andean passes.

Consider that officer of the South Moroccan Rifles who, during the war in the Rif, was in command of an outpost set down between two mountains filled with enemy tribesmen. One day, down from the mountain to the west came a group seeking a parley. Arabs and Frenchmen were talking over their tea when of a sudden a volley rang out. The tribesmen from the other mountain were charging the post. When the commandant sought to dismiss his guests before fighting off their allies, they said to him: "Today we are your guests. God will not allow us to desert you." They fought beside his men, saved the post, and then climbed back into their eyrie.

But on the eve of the day when their turn had come to pounce upon the post they sent again to the commandant.

"We came to your aid the other day," their chief said.

"True."

"We used up three hundred of our cartridges for you."

"Very likely."

"It would be only just that you replace them for us."

The commandant was an officer and a gentleman. They were given their cartridges.

Truth, for any man, is that which makes him a man. A man who has fraternized with men on this high plane, who has displayed this sportsmanship and has seen the rules of the game so nobly observed on both sides in matters of life and death, is obviously not to be mentioned in the same breath with the shabby hearty demagogue who would have expressed his fraternity with the Arabs by a great clap on the shoulders and a spate of flattering words that would have humiliated them. You might argue with the captain that all was fair in war, but if you did he would feel a certain pitying contempt for you. And he would be right.

Meanwhile, you are equally right to hate war.

If our purpose is to understand mankind and its yearnings, to grasp the essential reality of mankind, we must never set one man's truth against another's. All beliefs are demonstrably true. All men are demonstrably in the right. Anything can be demonstrated by logic. I say that that man is right who blames all the ills of the world upon hunchbacks. Let us declare war on hunchbacks—and in the twinkling of an eye all of us will hate them fanatically. All of us will join to avenge the crimes of the hunchbacks. Assuredly, hunchbacks, too, do commit crimes.

But if we are to succeed in grasping what is essential in man, we must put aside the passions that divide us and that, once they are accepted, sow in the wind a whole Koran of unassailable verities and fanaticisms. Nothing is easier than to divide men into rightists and leftists, hunchbacks and straightbacks, fascists and democrats—and these distinctions will be perfectly just. But truth, we know, is that which clarifies, not that which confuses. Truth is the language that expresses universality. Newton did not "discover" a law that lay hidden from man like the answer to a rebus. He accomplished a creative operation. He founded a human speech which could express at one and the same time the fall of an apple and the rising of the sun. Truth is not that which is demonstrable but that which is ineluctable.

There is no profit in discussing ideologies. If all of them are logically demonstrable then all of them must contradict one other. To agree to discuss them is tantamount to despairing of the salvation of mankind—whereas everywhere about us men manifest identical yearnings.

What all of us want is to be set free. The man who sinks his pickaxe into the ground wants that stroke to mean something. The convict's stroke is not the same as the prospector's, for the obvious reason that the prospector's stroke has meaning and the convict's stroke has none. It would be a mistake to think that the prison exists at the point where the convict's stroke is dealt. Prison is not a mere physical horror. It is using a pickaxe to no purpose that makes a prison; the horror resides in the failure to enlist all those who swing the pick in the community of mankind.

We all yearn to escape from prison.

There are two hundred million men in Europe whose existence has no meaning and who yearn to come alive. Industry has torn them from the idiom of their peasant lineage and has locked them up in those enormous ghettos that are like railway yards heaped with blackened trucks. Out of the depth of their slums these men yearn to be awakened. There are others, caught in the wheels of a thousand trades, who are forbidden to share in the joys known to a Mermoz, to a priest, to a man of science. Once it was believed that to bring these creatures to manhood it was enough to feed them, clothe them, and look to their everyday needs; but we see now that the result of this has been to turn out petty shopkeepers, village politicians, hollow technicians devoid of an inner life. Some indeed were well taught, but no one troubled to cultivate any of them. People who believe that culture consists in the capacity to remember formulae have a paltry notion of what it is. Of course any science student can tell us more about Nature and her laws than can Descartes or Newton,—but what can he tell us about the human spirit?

With more or less awareness, all men feel the need to come alive. But most of the methods suggested for bringing this about are snares and delusions. Men can of course be stirred into life by being dressed up in uniforms and made to blare out chants of war. It must be confessed that this is one way for men to break bread with comrades and to find what they are seeking, which is a sense of something universal, of self-fulfilment. But of this bread men die.

It is easy to dig up wooden idols and revive ancient and more or less workable myths like Pan-Germanism or the Roman Empire. The Germans can intoxicate themselves with the intoxication of being Germans and compatriots of Beethoven. A stoker in the hold of a freighter can be made drunk with this drink. What is more difficult is to bring up a Beethoven out of the stoke-hold.

These idols, in sum, are carnivorous idols. The man who dies for the progress of science or the healing of the sick serves life in his very dying. It may be glorious to die for the expansion of territory, but modern warfare destroys what it claims to foster. The day is gone when men sent life coursing through the veins of a race by the sacrifice of a little blood. War carried on by gas and bombing is no longer war, it is a kind of bloody surgery. Each side settles down behind a concrete wall and finds nothing better to do than to send forth, night after night, squadrons of planes to bomb the guts of the other side, blow up its factories, paralyze its production, and abolish its trade. Such a war is won by him who rots last—but in the end both rot together.

In a world become a desert we thirst for comradeship. It is the savor of bread broken with comrades that makes us accept the values of war. But there are other ways than war to bring us the warmth of a race, shoulder to shoulder, towards an identical goal. War has tricked us. It is not true that hatred adds anything to the exaltation of the race.

Why should we hate one another? We all live in the same cause, are borne through life on the same planet, form the crew of the same ship. Civilizations may, indeed, compete to bring forth new syntheses, but it is monstrous that they should devour one another.

To set man free it is enough that we help one another to realize that there does exist a goal towards which all mankind is striving. Why should we not strive towards that goal together, since it is what unites us all? The surgeon pays no heed to the moanings of his patient: beyond that pain it is man he is seeking to heal. That surgeon speaks a universal language. The physicist does the same when he ponders those almost divine equations in which he seizes the whole physical universe from the atom to the nebula. Even the simple shepherd modestly watching his sheep under the stars would discover, once he understood the part he was playing, that he was something more than a servant, was a sentinel. And each sentinel among men is responsible for the whole of the empire.

It is impossible not to believe that the shepherd wants to understand. One day, on the Madrid front, I chanced upon a school that stood on a hill surrounded by a low stone wall some five hundred yards behind the trenches. A corporal was teaching botany that day. He was lecturing on the fragile organs of a poppy held in his hands. Out of the surrounding mud, and in spite of the wandering shells that dropped all about, he had drawn like a magnet an audience of stubble-bearded soldiers who squatted tailor fashion and listened with their chins in their hands to a discourse of which they understood not a word in five. Something within them had said: "You are but brutes fresh from your caves. Go along! Catch up with humanity!" And they had hurried on their muddy clogs to overtake it.

It is only when we become conscious of our part in life, however modest, that we shall be happy. Only then will we be able to live in peace and die in peace, for only this lends meaning to life and to death.

Death is sweet when it comes in its time and in its place, when it is part of the order of things, when the old peasant of Provence, at the end of his reign, remits into the hands of his sons his parcel of goats and olive-trees in order that they in their turn transmit them to their sons. When one is part of a peasant lineage, one's death is only half a death. Each life in turn bursts like a pod and sends forth its seed.

I stood once with three peasants in the presence of their dead mother. Sorrow filled the room. For a second time, the umbilical cord had been cut. For a second time the knot had been loosed, the knot that bound one generation to another. Of a sudden the three sons had felt themselves alone on earth with everything still to be learned. The magnetic pole round which they had lived was gone; their mother's table, where they had collected on feast days with their families, was no more. But I could see in this rupture that it was possible for life to be granted a second time. Each of these sons was now to be the head of a family, was to be a rallying point and a patriarch, until that day when each would pass on the staff of office to the brood of children now murmuring in the courtyard.

I looked at their mother, at the old peasant with the firm peaceful face, the tight lips, the human face transformed into a stone mask. I saw in it the faces of her sons. That mask had served to mould theirs. That body had served to mould the bodies of these three exemplary men who stood there as upright as trees. And now she lay broken but at rest, a vein from which the gold had been extracted. In their turn, her sons and daughters would bring forth men from their mould. One does not die on a farm: their mother is dead, long live their mother!

Sorrowful, yes, but so simple was this image of a lineage dropping one by one its white-haired members as it made its way through time and through its metamorphoses towards a truth that was its own.

That same day, when the tocsin tolled to announce to the countryside the death of this old woman, it seemed to me not a song of despair but a discreet and tender chant of joy. In that same voice the church-bell celebrated birth and death, christening and burial, the passage from one generation to the next. I was suffused with a gentle peace of soul at this sound which announced the betrothal of a poor old woman and the earth.

This was life that was handed on here from generation to generation with the slow progress of a tree's growth, but it was also fulfilment. What a mysterious ascension! From a little bubbling lava, from the vague pulp of a star, from a living cell miraculously fertilized, we have issued forth and have bit by bit raised ourselves to the writing of cantatas and the weighing of nebulae.

This peasant mother had done more than transmit life, she had taught her sons a language, had handed on to them the lot so slowly garnered through the centuries, the spiritual patrimony of traditions, concepts, and myths that make up the whole of the difference between Newton or Shakespeare and the caveman.

What we feel when we are hungry, when we feel that hunger which drew the Spanish soldiers under fire towards that botany lesson, drew Mermoz across the South Atlantic, draws a man to a poem, is that the birth of man is not yet accomplished, that we must take stock of ourselves and our universe. We must send forth pontoons into the night. There are men unaware of this, imagining themselves wise and self-regarding because they are indifferent. But everything in the world gives the lie to their wisdom.

Comrades of the air! I call upon you to bear me witness. When have we felt ourselves happy men?

# X

## *Conclusion*

Here, in the final pages of this book, I remember again those musty civil servants who served as our escort in the omnibus when we set out to fly our first mails, when we prepared ourselves to be transformed into men—we who had had the luck to be called. Those clerks were kneaded of the same stuff as the rest of us, but they knew not that they were hungry.

To come to man's estate it is not necessary to get oneself killed round Madrid, or to fly mail planes, or to struggle wearily in the snows out of respect for the dignity of life. The man who can see the miraculous in a poem, who can take pure joy from music, who can break his bread with comrades, opens his window to the same refreshing wind off the sea. He too learns a language of men.

But too many men are left unawakened.

A few years ago, in the course of a long railway journey, I was suddenly seized by a desire to make a tour of the little country in which I was locked up for three days, cradled in that rattle that is like the sound of pebbles rolled over and over by the waves; and I got up out of my berth. At one in the morning I went through the train in all its length. The sleeping cars were empty. The first-class carriages were empty. They put me in mind of the luxurious hotels on the Riviera that open in winter for a single guest, the last representative of an extinct fauna. A sign of bitter times.

But the third-class carriages were crowded with hundreds of Polish workmen sent home from France. I made my way along those passages, stepping over sprawling bodies and peering into the carriages. In the dim glow cast by the night-lamps into these barren and comfortless compartments I saw a confused mass of people churned about by the swaying of the train, the whole thing looking and smelling like a barrack-room. A whole nation returning to its native poverty seemed to sprawl there in a sea of bad dreams. Great shaven heads rolled on the cushionless benches. Men, women, and children were stirring in their sleep, tossing from left to right and back again as if attacked by all the noises and jerkings that threatened them in their oblivion. They had not found the hospitality of a sweet slumber.

Looking at them I said to myself that they had lost half their human quality. These people had been knocked about from one end of Europe to the other by the economic currents; they had been torn from their little houses in the north of France, from their tiny garden-plots, their three pots of geranium that always stood in the windows of the Polish miners' families. I saw lying beside them pot and pans, blankets, curtains, bound into bundles badly tied and swollen with hernias.

Out of all that they had caressed or loved in France, out of everything they had succeeded in taming in their four or five years in my country—the cat, the dog, the geranium—they had been able to bring away with them only a few kitchen utensils, two or three blankets, a curtain or so.

A baby lay at the breast of a mother so weary that she seemed asleep. Life was being transmitted in the shabbiness and the disorder of this journey. I looked at the father. A powerful skull as naked as a stone. A body hunched over in uncomfortable sleep, imprisoned in working clothes, all humps and hollows. The man looked like a lump of clay, like one of those sluggish and shapeless derelicts that crumple into sleep in our public markets.

And I thought: The problem does not reside in this poverty, in this filth, in this ugliness. But this same man and this same woman met one day. This man must have smiled at this woman. He may, after his work was done, have brought her flowers. Timid and awkward, perhaps he trembled lest she disdain him. And this woman, out of natural coquetry, this woman sure of her charms, perhaps took pleasure in teasing him. And this man, this man who is now no more than a machine for swinging a pick or a sledgehammer, must have felt in his heart a delicious anguish. The mystery is that they should have become these lumps of clay. Into what terrible mould were they forced? What was it that marked them like this as if they had been put through a monstrous stamping machine? A deer, a gazelle, any animal grown old, preserves its grace. What is it that corrupts this wonderful clay of which man is kneaded?

I went on through these people whose slumber was as sinister as a den of evil. A vague noise floated in the air made up of raucous snores, obscure moanings, and the scraping of clogs as their wearers, broken on one side, sought comfort on the other. And always the muted accompaniment of those pebbles rolled over and over by the waves.

I sat down face to face with one couple. Between the man and the woman a child had hollowed himself out a place and fallen asleep. He turned in his slumber, and in the dim lamplight I saw his face. What an adorable face! A golden fruit had been born of these two peasants. Forth from this sluggish scum had sprung this miracle of delight and grace.

I bent over the smooth brow, over those mildly pouting lips, and I said to myself: This is a musician's face. This is the child Mozart. This is a life full of beautiful promise. Little princes in legends are not different from this. Protected, sheltered, cultivated, what could not this child become?

When by mutation a new rose is born in a garden, all the gardeners rejoice. They isolate the rose, tend it, foster it. But there is no gardener for men. This little Mozart will be shaped like the rest by the common stamping machine. This little Mozart will love shoddy music in the stench of night dives. This little Mozart is condemned.

I went back to my sleeping car. I said to myself: Their fate causes these people no suffering. It is not an impulse to charity that has upset me like this. I am not weeping over an eternally open wound. Those who carry the wound do not feel it. It is the human race and not the individual that is wounded here, is outraged here. I do not believe in pity. What torments me tonight is the gardener's point of view. What torments me is not this poverty to which after all a man can accustom himself as easily as to sloth. Generations of Orientals live in filth and love it. What torments me is not the humps nor the hollows nor the ugliness. It is the sight, a little bit in all these men, of Mozart murdered.

Only the Spirit, if it breathe upon the clay, can create Man.

# NIGHT FLIGHT

*Translated from the French*
*By Stuart Gilbert*

# *Preface*

The *sine qua non* for the air-line companies was to compete in speed with all other systems of transport. In the course of this book Rivière, that leader to the manner born, sums up the issues. "It is a matter of life and death for us; for the lead we gain by day on ships and railways is lost each night." This night service—much criticized at the start but subsequently, once the experimental stage was over, accepted as a practical proposition—still involved at the time of this narrative considerable risks. For to the impalpable perils of all air-routes and their manifold surprises accrued the night's dark treachery. I hasten to add that, great though these risks still are, they are growing daily less, for each successive trip facilitates and improves the prospects of the next one. Aviation, like the exploration of uncharted lands, has its early heroic age and "Night Flight," which describes the tragic adventure of one of these pioneers of the air, sounds, naturally enough, the authentic epic note.

The hero of "Night Flight," though human through and through, rises to superhuman heights of valor. The quality which I think delights one most of all in this stirring narrative is its nobility. Too well we know man's failings, his cowardice and lapses, and our writers of to-day are only too proficient in exposing these; but we stood in need of one to tell us how a man may be lifted far above himself by his sheer force of will.

More striking even than the aviator himself is, in my opinion, Rivière, his chief. The latter does not act, himself; he impels to action, breathes into his pilots his own virtue and exacts the utmost from them, constraining them to dare greatly. His iron will admits no flinching, and the least lapse is punished by him. At first sight his severity may seem inhuman and excessive. But its target is not the man himself, whom Rivière aspires to mold, but the man's blemishes. In his portrayal of this character we feel the author's profound admiration. I am especially grateful to him for bringing out a paradoxical truth which seems to me of great psychological import; that man's happiness lies not in freedom but in his acceptance of a duty. Each of the characters in this book is wholeheartedly, passionately devoted to that which duty bids him do, and it is in fulfilling this perilous task, and only thus, that he attains contentedness and peace. Reading between the lines we discover that Rivière is anything but insensitive (the narrative of his interview with the wife of the lost pilot is infinitely touching) and he needs quite as much courage to give his orders as the pilots need to carry them out.

"To make oneself beloved," he says, "one need only show pity. I show little pity, or I hide it. . . . My power sometimes amazes me." And, again: "Love the men under your orders, but do not let them know it."

A sense of duty commands Rivière in all things, "the dark sense of duty, greater than that of love." Man is not to seek an end within himself but to submit and sacrifice his all to some strange thing that commands him and lives through him. It pleases me here to find that selfsame "dark sense" which inspired my Prometheus to his paradox: "Man I love not; I love that which devours him." This is the mainspring of every act of heroism. " 'We behave,' thought Rivière, 'as if there were something of higher value than human life. . . . But what thing?' " And again: "There is perhaps something else, something more lasting, to be saved; and perhaps it was to save this part of man that Rivière was working." A true saying.

In an age when the idea of heroism seems likely to quit the army, since manly virtues may play no part in those future wars whose horrors are foreshadowed by our scientists, does not aviation provide the most admirable and worthy field for the display of prowess? What would otherwise be rashness ceases to be such when it is part and parcel of an allotted task. The pilot who is forever risking his life may well smile at the current meaning we give to "courage." I trust that Saint-Exupéry will permit me to quote an old letter of his dating from the time when he was flying on the Casablanca-Dakar air-route.

"I don't know when I shall be back, I have had so much to do for several months, searches for lost airmen, salvage of planes that have come down in hostile territory, and some flights with the Dakar mail.

"I have just pulled off a little exploit; spent two days and nights with eleven Moors and a mechanic, salving a plane. Alarums and excursions, varied and impressive. I heard bullets whizzing over my head for the first time. So now I know how I behave under such conditions; much more calmly than the Moors. But I also came to understand something which had always puzzled me—why Plato (Aristotle?) places courage in the last degree of virtues. It's a concoction of feelings that are not so very admirable. A touch of anger, a spice of vanity, a lot of obstinancy and a tawdry 'sporting' thrill. Above all, a stimulation of one's physical energies, which, however, is oddly out of place. One just folds one's arms, taking deep breaths, across one's opened shirt. Rather a pleasant feeling. When it happens at night another feeling creeps into it—of having done something immensely silly. I shall never again admire a merely brave man."

By way of epigraph I might append to this quotation an aphorism from Quinton's book (which, however, I cannot commend without reserve). "A man keeps, like his love, his courage dark." Or, better still: "Brave men hide their deeds as decent folk their alms. They disguise them or make excuses for them."

Saint-Exupéry in all he tells us speaks as one who has "been through it." His personal contact with ever-recurrent danger seasons his book with an authentic and inimitable tang. We have had many stories of the War or of imaginary adventures which, if they showed the author as a man of nimble wit, brought smiles to the faces of such old soldiers or genuine adventurers as read them. I admire this work not only on its literary merits but for its value as a record of realities, and it is the unlikely combination of these two qualities which gives "Night Flight" its quite exceptional importance.

André Gide

# I

Already, beneath him, through the golden evening, the shadowed hills had dug their furrows and the plains grew luminous with long-enduring light. For in these lands the ground gives off this golden glow persistently, just as, even when winter goes, the whiteness of the snow persists.

Fabien, the pilot bringing the Patagonia air-mail from the far south to Buenos Aires, could mark night coming on by certain signs that called to mind the waters of a harbor—a calm expanse beneath, faintly rippled by the lazy clouds—and he seemed to be entering a vast anchorage, an immensity of blessedness.

Or else he might have fancied he was taking a quiet walk in the calm of evening, almost like a shepherd. The Patagonian shepherds move, unhurried, from one flock to another; and he, too, moved from one town to another, the shepherd of those little towns. Every two hours he met another of them, drinking at its riverside or browsing on its plain.

Sometimes, after a hundred miles of steppes as desolate as the sea, he encountered a lonely farm-house that seemed to be sailing backwards from him in a great prairie sea, with its freight of human lives; and he saluted with his wings this passing ship.

"San Julian in sight. In ten minutes we shall land."

The wireless operator gave their position to all the stations on the line. From Magellan Strait to Buenos Aires the airports were strung out across fifteen hundred miles and more, but this one led toward the frontiers of night, just as in Africa the last conquered hamlet opens on to the unknown.

The wireless operator handed the pilot a slip of paper: "There are so many storms about that the discharges are fouling my earphones. Shall we stop the night at San Julian?"

Fabien smiled; the sky was calm as an aquarium and all the stations ahead were signaling, *Clear sky: no wind.*

"No, we'll go on."

But the wireless operator was thinking: these storms had lodged themselves somewhere or other, as worms do in a fruit; a fine night, but they would ruin it, and he loathed entering this shadow that was ripe to rottenness.

As he slowed down his engine for the San Julian landing, Fabien knew that he was tired. All that endeared his life to man was looming up to meet him; men's houses, friendly little cafés, trees under which they walk. He was like some conqueror who, in the aftermath of victory, bends down upon his territories and now perceives the humble happiness of men. A need came over Fabien to lay his weapons down and feel the aching burden of his limbs—for even our misfortunes are a part of our belongings—and to stay, a simple dweller here, watching from his window a scene that would never change. This tiny village, he could gladly have made friends with it; the choice once made, a man accepts the issue of his venture and can love the life. Like love, it hems him in. Fabien would have wished to live a long while here—here to possess his morsel of eternity. These little towns, where he lived an hour, their gardens girdled by old walls over which he passed, seemed something apart and everlasting. Now the village was rising to meet the plane, opening out toward him. And there, he mused, were friendliness and gentle girls, white napery spread in quiet homes; all that is slowly shaped toward eternity. The village streamed past beneath his wings, yielding the secrets of closed gardens that their walls no longer guarded. He landed; and now he knew that he had seen nothing at all, only a few men slowly moving amongst their stones. The village kept, by its mere immobility, the secret of its passions and withheld its kindly charm; for, to master that, he would have needed to give up an active life.

The ten minutes' halt was ended and Fabien resumed his flight. He glanced back toward San Julian; all he now could see was a cluster of lights, then stars, then twinkling star-dust that vanished, tempting him for the last time.

"I can't see the dials; I'll light up."

He touched the switches, but the red light falling from the cockpit lamps upon the dial-hands was so diluted with the blue evening glow that they did not catch its color. When he passed his fingers close before a bulb, they were hardly tinged at all.

"Too soon."

But night was rising like a tawny smoke and already the valleys were brimming over with it. No longer were they distinguishable from the plains. The villages were lighting up, constellations that greeted each other across the dusk. And, at a touch of his finger, his flying-lights flashed back a greeting to them. The earth grew spangled with light-signals as each house lit its star, searching the vastness of the night as a lighthouse sweeps the sea. Now every place that sheltered human life was sparkling. And it rejoiced him to enter into this one night with a measured slowness, as into an anchorage.

He bent down into the cockpit; the luminous dial-hands were beginning to show up. The pilot read their figures one by one; all was going well. He felt at ease up here, snugly ensconced. He passed his fingers along a steel rib and felt the stream of life that flowed in it; the metal did not vibrate, yet it was alive. The engine's five-hundred horse-power bred in its texture a very gentle current, fraying its ice-cold rind into a velvety bloom. Once again the pilot in full flight experienced neither giddiness nor any thrill; only the mystery of metal turned to living flesh.

So he had found his world again. . . . A few digs of his elbow, and he was quite at home. He tapped the dashboard, touched the contacts one by one, shifting his limbs a little, and, settling himself more solidly, felt for the best position whence to gage the faintest lurch of his five tons of metal, jostled by the heaving darkness. Groping with his fingers, he plugged in his emergency-lamp, let go of it, felt for it again, made sure it held; then lightly touched each switch, to be certain of finding it later, training his hands to function in a blind man's world. Now that his hands had learnt their rôle by heart, he ventured to turn on a lamp, making the cockpit bright with polished fittings and then, as on a submarine about to dive, watched his passage into night upon the dials only. Nothing shook or rattled, neither gyroscope nor altimeter flickered in the least, the engine was running smoothly; so now he relaxed his limbs a little, let his neck sink back into the leather padding and fell into the deeply meditative mood of flight, mellow with inexplicable hopes.

Now, a watchman from the heart of night, he learnt how night betrays man's presence, his voices, lights, and his unrest. That star down there in the shadows, alone; a lonely house. Yonder a fading star; that house is closing in upon its love. . . . Or on its lassitude. A house that has ceased to flash its signal to the world. Gathered round their lamp-lit table, those peasants do not know the measure of their hopes; they do not guess that their desire carries so far, out into the vastness of the night that hems them in. But Fabien has met it on his path, when, coming from a thousand miles away, he feels the heavy ground-swell raise his panting plane and let it sink, when he has crossed a dozen storms like lands at war, between them neutral tracts of moonlight, to reach at last those lights, one following the other—and knows himself a conqueror. They think, these peasants, that their lamp shines only for that little table; but, from fifty miles away, some one has felt the summons of their light, as though it were a desperate signal from some lonely island, flashed by shipwrecked men toward the sea.

## II

Thus the three planes of the air-mail service, from Patagonia, Chile, and Paraguay, were converging from south, west, and north on Buenos Aires. Their arrival with the mails would give the signal for the departure, about midnight, of the Europe postal plane.

Three pilots, each behind a cowling heavy as a river-barge, intent upon his flight, were hastening through the distant darkness, soon to come slowly down, from a sky of storm or calm, like wild, outlandish peasants descending from their highlands.

Rivière, who was responsible for the entire service, was pacing to and fro on the Buenos Aires landing-ground. He was in silent mood, for, till the three planes had come in, he could not shake off a feeling of apprehension which had been haunting him all day. Minute by minute, as the telegrams were passed to him, Rivière felt that he had scored another point against fate, reduced the quantum of the unknown, and was drawing his charges in, out of the clutches of the night, toward their haven.

One of the hands came up to Rivière with a radio message.

"Chile mail reports: Buenos Aires in sight."

"Good."

Presently, then, Rivière would hear its drone; already the night was yielding up one of them, as a sea, heavy with its secrets and the cadence of the tides, surrenders to the shore a treasure long the plaything of the waves. And soon the night would give him back the other two.

Then to-day's work would be over. Worn out, the crews would go to sleep, fresh crews replace them. Rivière alone would have no respite; then, in its turn, the Europe mail would weigh upon his mind. And so it would always be. Always. For the first time in his life this veteran fighter caught himself feeling tired. Never could an arrival of the planes mean for him the victory that ends a war and preludes a spell of smiling peace. For him it meant just one more step, with a thousand more to follow, along a straight, unending road. Rivière felt as though for an eternity he had been carrying a crushing load on his uplifted arms; an endless, hopeless effort.

"I'm aging." If he no longer found a solace in work and work alone, surely he was growing old. He caught himself puzzling over problems which hitherto he had ignored. There surged within his mind, like a lost ocean, murmuring regrets, all the gentler joys of life that he had thrust aside. "Can it be coming on me—so soon?" He realized that he had always been postponing for his declining years, "when I have time for it," everything that makes life kind to men. As if it were ever possible to "have time for it" one day and realize at life's end that dream of peace and happiness! No, peace there could be none; nor any victory, perhaps. Never could all the air-mails land in one swoop once for all.

Rivière paused before Leroux; the old foreman was hard at work.

Leroux, too, had forty years of work behind him. All his energies were for his work. When at ten o'clock or midnight Leroux went home it certainly was not to find a change of scene, escape into another world. When Rivière smiled toward him, he raised his heavy head and pointed at a burnt-out axle. "Jammed it was, but I've fixed it up." Riviére bent down to look; duty had regained its hold upon him. "You should tell the shop to set them a bit looser." He passed his finger over the trace of seizing, then glanced again at Leroux. As his eyes lingered on the stern old wrinkled face, an odd question hovered on his lips and made him smile.

"Ever had much to do with love, Leroux, in your time?"

"Love, sir? Well, you see—"

"Hadn't the time for it, I suppose—like me."

"Not a great deal, sir."

Rivière strained his ears to hear if there were any bitterness in the reply; no, not a trace of it. This man, looking back on life, felt the quiet satisfaction of a carpenter who has made a good job of planing down a board: "There you are! *That's* done."

"There you are," thought Rivière. "My life's done."

Then, brushing aside the swarm of somber thoughts his weariness had brought, he walked toward the hangar; for the Chile plane was droning down toward it.

## III

The sound of the distant engine swelled and thickened; a sound of ripening. Lights flashed out. The red lamps on the light-tower silhouetted a hangar, radio standards, a square landing-ground. The setting of a gala night.

"There she comes!"

A sheaf of beams had caught the grounding plane, making it shine as if brand-new. No sooner had it come to rest before the hangar than mechanics and airdrome hands hurried up to unload the mail. Only Pellerin, the pilot, did not move.

"Well, aren't you going to get down?"

The pilot, intent on some mysterious task, did not deign to reply. Listening, perhaps, to sounds that he alone could hear, long echoes of the flight. Nodding reflectively, he bent down and tinkered with some unseen object. At last he turned toward the officials and his comrades, gravely taking stock of them as though of his possessions. He seemed to pass them in review, to weigh them, take their measure, saying to himself that he had earned his right to them, as to this hangar with its gala lights and solid concrete and, in the offing, the city, full of movement, warmth, and women. In the hollow of his large hands he seemed to hold this folk; they were his subjects, to touch or hear or curse, as the fancy took him. His impulse now was to curse them for a lazy crowd, so sure of life they seemed, gaping at the moon; but he decided to be genial instead.

". . . Drinks are on you!"

Then he climbed down.

He wanted to tell them about the trip.

"If only you knew . . . !"

Evidently, to his thinking, that summed it up, for now he walked off to change his flying gear.

As the car was taking him to Buenos Aires in the company of a morose inspector and Rivière in silent mood, Pellerin suddenly felt sad; of course, he thought, it's a fine thing for a fellow to have gone through it and, when he's got his footing again, let off a healthy volley of curses. Nothing finer in the world! But afterwards . . . when you look back on it all; you wonder, you aren't half so sure!

A struggle with a cyclone, that at least is a straight fight, it's *real*. But not that curious look things wear, the face they have when they think they are alone. His thoughts took form. "Like a revolution it is; men's faces turning only the least shade paler, yet utterly unlike themselves."

He bent his mind toward the memory.

He had been crossing peacefully the Cordillera of the Andes. A snow-bound stillness brooded on the ranges; the winter snow had brought its peace to all this vastness, as in dead castles the passing centuries spread peace. Two hundred miles without a man, a breath of life, a movement; only sheer peaks that, flying at twenty thousand feet, you almost graze, straight-falling cloaks of stone, an ominous tranquillity.

It had happened somewhere near the Tupungato Peak. . . .

He reflected. . . . Yes, it was there he saw a miracle take place.

For at first he had noticed nothing much, felt no more than a vague uneasiness—as when a man believes himself alone, but is not; some one is watching him. Too late, and how he could not comprehend, he realized that he was hemmed in by anger. Where was it coming from, this anger? What told him it was oozing from the stones, sweating from the snow? For nothing seemed on its way to him, no storm was lowering. And still—another world, like it and yet unlike, was issuing from the world around him. Now all those quiet-looking peaks, snow-caps, and ridges, growing faintly grayer, seemed to spring to life, a people of the snows. And an inexplicable anguish gripped his heart.

Instinctively he tightened his grasp on the controls. Something he did not understand was on its way and he tautened his muscles, like a beast about to spring. Yet, as far as eye could see, all was at peace. Peaceful, yes, but tense with some dark potency.

Suddenly all grew sharp; peaks and ridges seemed keen-edged prows cutting athwart a heavy head wind. Veering around him, they deployed like dreadnoughts taking their positions in a battle-line. Dust began to mingle with the air, rising and hovering, a veil above the snow. Looking back to see if retreat might still be feasible, he shuddered; all the Cordillera behind him was in seething ferment.

"I'm lost!"

On a peak ahead of him the snow swirled up into the air—a snow volcano. Upon his right flared up another peak and, one by one, all the summits grew lambent with gray fire, as if some unseen messenger had touched them into flame. Then the first squall broke and all the mountains round the pilot quivered.

Violent action leaves little trace behind it and he had no recollection of the gusts that buffeted him then from side to side. Only one clear memory remained; the battle in a welter of gray flames.

He pondered.

"A cyclone, that's nothing. A man just saves his skin! It's what comes before it—the thing one meets upon the way!"

But already, even as he thought he had recalled it, that one face in a thousand, he had forgotten what it was like.

## IV

Rivière glanced at the pilot. In twenty minutes Pellerin would step from the car, mingle with the crowd, and know the burden of his lassitude. Perhaps he would murmur: "Tired out as usual.

It's a dog's life!" To his wife he would, perhaps, let fall a word or two: "A fellow's better off here than flying above the Andes!" And yet that world to which men hold so strongly had almost slipped from him; he had come to know its wretchedness. He had returned from a few hours' life on the other side of the picture, ignoring if it would be possible for him ever to retrieve this city with its lights, ever to know again his little human frailties, irksome yet cherished childhood friends.

"In every crowd," Rivière mused, "are certain persons who seem just like the rest, yet they bear amazing messages. Unwittingly, no doubt, unless—" Rivière was chary of a certain type of admirers, blind to the higher side of this adventure, whose vain applause perverted its meaning, debased its human dignity. But Pellerin's inalienable greatness lay in this—his simple yet sure awareness of what the world, seen from a special angle, signified, his massive scorn of vulgar flattery. So Rivière congratulated him: "Well, how did you bring it off?" And loved him for his knack of only "talking shop," referring to his flight as a blacksmith to his anvil.

Pellerin began by telling how his retreat had been cut off. It was almost as if he were apologizing about it. "There was nothing else for it!" Then he had lost sight of everything, blinded by the snow. He owed his escape to the violent air-currents which had driven him up to twenty-five thousand feet. "I guess they held me all the way just above the level of the peaks." He mentioned his trouble with gyroscope and how he had had to shift the air-inlet, as the snow was clogging it; "forming a frost-glaze, you see." After that another set of air-currents had driven Pellerin down and, when he was only at ten thousand feet or so, he was puzzled why he had not run into anything. As a matter of fact he was already above the plains. "I spotted it all of a sudden when I came out into a clear patch." And he explained how it had felt at that moment; just as if he had escaped from a cave.

"Storm at Mendoza, too?"

"No. The sky was clear when I made my landing, not a breath of wind. But the storm was at my heels all right!"

It was such a damned queer business, he said; that was why he mentioned it. The summits were lost in snow at a great height while the lower slopes seemed to be streaming out across the plain, like a flood of black lava which swallowed up the villages one by one. "Never saw anything like it before. . . ." Then he relapsed into silence, gripped by some secret memory.

Rivière turned to the inspector.

"That's a Pacific cyclone; it's too late to take any action now. Anyhow these cyclones never cross the Andes."

No one could have foreseen that this particular cyclone would continue its advance toward the east.

The inspector, who had no ideas on the subject, assented.

The inspector seemed about to speak. Then he hesitated, turned toward Pellerin, and his Adam's apple stirred. But he held his peace and, after a moment's thought, resumed his air of melancholy dignity, looking straight before him.

That melancholy of his, he carried it about with him everywhere, like a handbag. No sooner had he landed in Argentina than Rivière had appointed him to certain vague functions, and now his large hands and inspectorial dignity got always in his way. He had no right to admire imagination or ready wit; it was his business to commend punctuality and punctuality alone. He had no right to take a glass of wine in company, to call a comrade by his Christian name or risk a joke; unless, of course, by some rare chance, he came across another inspector on the same run.

"It's hard luck," he thought, "always having to be a judge."

As a matter of fact he never judged; he merely wagged his head. To mask his utter ignorance he would slowly, thoughtfully, wag his head at everything that came his way, a movement that struck fear into uneasy consciences and ensured the proper upkeep of the plant.

He was not beloved—but then inspectors are not made for love and such delights, only for drawing up reports. He had desisted from proposing changes of system or technical improvements since Rivière had written: *Inspector Robineau is requested to supply reports, not poems. He will be putting his talents to better use by speeding up the personnel.* From that day forth Inspector Robineau had battened on human frailties, as on his daily bread; on the mechanic who had a glass too much, the airport overseer who stayed up of nights, the pilot who bumped a landing.

Rivière said of him: "He is far from intelligent, but very useful to us, such as he is." One of the rules which Rivière rigorously imposed—upon himself—was a knowledge of his men. For Robineau the only knowledge that counted was knowledge of the *orders*.

"Robineau," Rivière had said one day, "you must cut the punctuality bonus whenever a plane starts late."

"Even when it's nobody's fault? In case of fog, for instance?"

"Even in case of fog."

Robineau felt a thrill of pride in knowing that his chief was strong enough not to shrink from being unjust. Surely Robineau himself would win reflected majesty from such overweening power!

"You postponed the start till six fifteen," he would say to the airport superintendents. "We cannot allow your bonus."

"But, Monsieur Robineau, at five thirty one couldn't see ten yards ahead!"

"Those are the *orders*."

"But, Monsieur Robineau, we couldn't sweep the fog away with a broom!"

He alone amongst all these nonentities knew the secret; if you only punish men enough, the weather will improve!

"He never thinks at all," said Rivière of him, "and that prevents him from thinking wrong."

The pilot who damaged a plane lost his no-accident bonus.

"But supposing his engine gives out when he is over a wood?" Robineau inquired of his chief.

"Even when it occurs above a wood."

Robineau took to heart the *ipse dixit.*

"I regret," he would inform the pilots with cheerful zest, "I regret it very much indeed, but you should have had your breakdown somewhere else."

"But, Monsieur Robineau, one doesn't choose the place to have it."

"Those are the orders."

The orders, thought Rivière, are like the rites of a religion; they may look absurd but they shape men in their mold. It was no concern to Rivière whether he seemed just or unjust. Perhaps the words were meaningless to him. The little townsfolk of the little towns promenade each evening round a bandstand and Rivière thought: It's nonsense to talk of being just or unjust toward them; they don't exist.

For him, a man was a mere lump of wax to be kneaded into shape. It was his task to furnish this dead matter with a soul, to inject will-power into it. Not that he wished to make slaves of his men; his aim was to raise them above themselves. In punishing them for each delay he acted, no doubt, unjustly, but he bent the will of every crew to punctual departure; or, rather, he bred in them the will to keep to time. Denying his men the right to welcome foggy weather as the pretext for a leisure hour, he kept them so breathlessly eager for the fog to lift that even the humblest mechanic felt a twinge of shame for the delay. Thus they were quick to profit by the least rift in the armor of the skies.

"An opening on the north; let's be off!"

Thanks to Rivière the service of the mails was paramount over twenty thousand miles of land and sea.

"The men are happy," he would say, "because they like their work, and they like it because I am hard."

And hard he may have been—still he gave his men keen pleasure for all that. "They need," he would say to himself, "to be urged on toward a hardy life, with its sufferings and its joys; only that matters."

As the car approached the city, Rivière instructed the driver to take him to the Head Office. Presently Robineau found himself alone with Pellerin and a question shaped itself upon his lips.

## V

Robineau was feeling tired to-night. Looking at Pellerin—Pellerin the Conqueror—he had just discovered that his own life was a gray one. Worst of all, he was coming to realize that, for all his rank of inspector and authority, he, Robineau, cut a poor figure beside this travel-stained and weary pilot, crouching in a corner of the car, his eyes closed and hands all grimed with oil. For the first time, Robineau was learning to admire. A need to speak of this came over him and, above all, to make a friend.

He was tired of his journey and the day's rebuffs and felt perhaps a little ridiculous. That very evening, when verifying the gasoline reserve, he had botched his figures and the agent, whom he had wanted to catch out, had taken compassion and totted them up for him. What was worse, he had commented on the fitting of a Model B.6 oil-pump, mistaking it for the B.4 type, and the mechanics with ironic smiles had let him maunder on for twenty minutes about this "inexcusable stupidity"—his own stupidity.

He dreaded his room at the hotel. From Toulouse to Buenos Aires, straight to his room he always went once the day's work was over. Safely ensconced and darkly conscious of the secrets he carried in his breast, he would draw from his bag a sheet of paper and slowly inscribe *Report* on it, write a line or two at random, then tear it up. He would have liked to save the company from some tremendous peril; but it was not in any danger. All he had saved so far was a slightly rusted propeller-boss. He had slowly passed his finger over the rust with a mournful air, eyed by an airport overseer, whose only comment was: "Better call up the last halt; this plane's only just in." Robineau was losing confidence in himself.

At a venture he essayed a friendly move. "Would you care to dine with me?" he asked Pellerin. "I'd enjoy a quiet chat; my job's pretty exhausting at times."

Then, reluctant to quit his pedestal too soon, he added: "The responsibility, you know."

His subordinates did not much relish the idea of intimacy with Robineau; it had its dangers. "If he's not dug up something for his report, with an appetite like his, I guess he'll just eat me up!"

But Robineau's mind this evening was full of his personal afflictions. He suffered from an annoying eczema, his only real secret; he would have liked to talk about his trouble, to be pitied and, now that pride had played him false, find solace in humility. Then again there was his mistress over there in France, who had to hear the nightly tale of his inspections whenever he returned. He hoped to impress her thus and earn her love—his usual luck!—he only seemed to aggravate her. He wanted to talk about her, too.

"So you'll come to dinner?"

Good-naturedly Pellerin assented.

## VI

The clerks were drowsing in the Buenos Aires office when Rivière entered. He had kept his overcoat and hat on, like the incessant traveler he always seemed to be. His spare person took up so little room, his clothes and graying hair so aptly fitted into any scene, that when he went by hardly any one noticed it. Yet, at his entry, a wave of energy traversed the office. The staff bustled, the head clerk hurriedly compiled the papers remaining on his desk, typewriters began to click.

The telephonist was busily slipping his plugs into the standard and noting the telegrams in a bulky register. Rivière sat down and read them.

All that he read, the Chile episode excepted, told of one of those favored days when things go right of themselves and each successive message from the airports is another bulletin of victory. The Patagonia mail, too, was making headway; all the planes were ahead of time, for fair winds were bearing them northward on a favoring tide.

"Give me the weather reports."

Each airport vaunted its fine weather, clear sky, and clement breeze. The mantle of a golden evening had fallen on South America. And Rivière welcomed this friendliness of things. True, one of the planes was battling somewhere with the perils of the night, but the odds were in its favor.

Rivière pushed the book aside.

"That will do."

Then, a night-warden whose charge was half the world, he went out to inspect the men on night duty, and came back.

Later, standing at an open window, he took the measure of the darkness. It contained Buenos Aires yonder, but also, like the hull of some huge ship, America. He did not wonder at this feeling of immensity; the sky of Santiago de Chile might be a foreign sky, but once the air-mail was in flight toward Santiago you lived, from end to journey's end, under the same dark vault of heaven. Even now the Patagonian fishermen were gazing at the navigation lights of the plane whose messages were being awaited here. The vague unrest of an aëroplane in flight brooded not only on Rivière's heart but, with the droning of the engine, upon the capitals and little towns.

Glad of this night that promised so well, he recalled those other nights of chaos, when a plane had seemed hemmed in with dangers, its rescue well-nigh a forlorn hope, and how to the Buenos Aires Radio Post its desperate calls came faltering through, fused with the atmospherics of the storm. Under the leaden weight of sky the golden music of the waves was tarnished. Lament in the minor of a plane sped arrowwise against the blinding barriers of darkness, no sadder sound than this!

Rivière remembered that the place of an inspector, when the staff is on night duty, is in the office.

"Send for Monsieur Robineau."

Robineau had all but made a friend of his guest, the pilot. Under his eyes he had unpacked his suitcase and revealed those trivial objects which link inspectors with the rest of men; some shirts in execrable taste, a dressing-set, the photograph of a lean woman, which the inspector pinned to the wall. Humbly thus he imparted to Pellerin his needs, affections, and regrets. Laying before the pilot's eyes his sorry treasures, he laid bare all his wretchedness. A moral eczema. His prison.

But a speck of light remained for Robineau, as for every man, and it was in a mood of quiet ecstasy that he drew, from the bottom of his valise, a little bag carefully wrapped up in paper. He fumbled with it some moments without speaking. Then he unclasped his hands.

"I brought this from the Sahara."

The inspector blushed to think that he had thus betrayed himself. For all his chagrins, domestic misadventures, for all the gray reality of life he had a solace, these little blackish pebbles—talismans to open doors of mystery.

His blush grew a little deeper. "You find exactly the same kind in Brazil."

Then Pellerin had slapped the shoulder of an inspector poring upon Atlantis and, as in duty bound, had asked a question.

"Keen on geology, eh?"

"Keen? I'm mad about it!"

All his life long only the stones had not been hard on him.

Hearing that he was wanted, Robineau felt sad but forthwith resumed his air of dignity.

"I must leave you. Monsieur Rivière needs my assistance for certain important problems."

When Robineau entered the office, Rivière had forgotten all

about him. He was musing before a wall-map on which the company's air-lines were traced in red. The inspector awaited his chief's orders. Long minutes passed before Rivière addressed him, without turning his head.

"What is your idea of this map, Robineau?"

He had a way of springing conundrums of this sort when he came out of a brown study.

"The map, Monsieur Rivière? Well—"

As a matter of fact he had no ideas on the subject; nevertheless, frowning at the map, he roved all Europe and America with an inspectorial eye. Meanwhile Rivière, in silence, pursued his train of thought. "On the face of it, a pretty scheme enough—but it's ruthless. When one thinks of all the lives, young fellows' lives, it has cost us! It's a fine, solid thing and we must bow to its authority, of course; but what a host of problems it presents!" With Rivière, however, nothing mattered save the end in view.

Robineau, standing beside him with his eyes fixed on the map, was gradually pulling himself together. Pity from Rivière was not to be expected; that he knew. Once he had chanced it, explaining how that grotesque infirmity of his had spoilt his life. All he had got from Rivière was a jeer. "Stops you sleeping, eh? So much the better for your work!"

Rivière spoke only half in jest. One of his sayings was: "If a composer suffers from loss of sleep and his sleeplessness induces him to turn out masterpieces, what a profitable loss it is!" One day, too, he had said of Leroux: "Just look at him! I call it a fine thing, ugliness like that—so perfect that it would warn off any sweetheart!" And perhaps, indeed, Leroux owed what was finest in him to his misfortune, which obliged him to live only for his work.

"Pellerin's a great friend of your, isn't he, Robineau?"

"Well—"

"I'm not reproaching you."

Rivière made a half-turn and with bowed head, taking short steps, paced to and fro with Robineau. A bitter smile, incomprehensible to Robineau, came to his lips.

"Only . . . . only you are his chief, you see."

"Yes," said Robineau.

Rivière was thinking how to-night, as every night, a battle was in progress in the southern sky. A moment's weakening of the will might spell defeat; there was, perhaps, much fighting to be done before the dawn.

"You should keep your place, Robineau." Rivière weighed his words. "You may have to order this pilot to-morrow night to start on a dangerous flight. He will have to obey you."

"Yes."

"The lives of men worth more than you are in your hands." He seemed to hesitate. "It's a serious matter."

For a while Rivière paced the room in silence, taking his little steps.

"If they obey you because they like you, Robineau, you're fooling them. You have no right to ask any sacrifice of them."

"No, of course not."

"And if they think that your friendship will get them off disagreeable duties, you're fooling them again. They have to obey in any case. Sit down."

With a touch of his hand Rivière gently propelled Inspector Robineau toward the desk.

"I am going to teach you a lesson, Robineau. If you feel run down it's not these men's business to give you energy. You are their chief. Your weakness is absurd. Now write!"

"I—"

"Write. *Inspector Robineau imposes the penalty stated hereunder on Pellerin, Pilot, on the following grounds. . . .* You will discover something to fill in the blanks."

"Sir!"

"Act as though you understood, Robineau. Love the men under your orders—but do not let them know it."

So, once more, Robineau would supervise the cleaning of each propeller-boss, with zest.

An emergency landing-ground sent in a radio message. *Plane in sight. Plane signals: Engine Trouble; about to land.*

That meant half an hour lost. Rivière felt that mood of irritation the traveler knows when his express is held up by a signal and the minutes no longer yield their toll of passing hedgerows. The large clock-hand was turning now an empty hemicycle, within whose compass so many things might have fitted in. To while away the interval Rivière went out and now the night seemed hollow as a stage without an actor. Wasted—a night like this! He nursed a grudge against that cloudless sky with its wealth of stars, the moon's celestial beacon, the squandered gold of such a night. . . .

But, once the plane had taken off, the night once more grew full of beauty and enthralment; for now the womb of night was carrying life, and over it Rivière kept his watch.

"What weather have you?"

He had the query transmitted to the crew. Ten seconds later the reply came in: "Very fine."

There followed a string of names, towns over which the plane had passed and, for Rivière's ears, these were so many names of cities falling one by one before a conqueror.

# VII

An hour later the wireless operator on the Patagonia mail felt himself gently lifted as though some one were tugging at his shoulder. He looked around; heavy clouds were putting out the stars. He leaned toward the earth, trying to see the village lights, shining like glowworms in the grass, but in those fields of darkness no light sparkled.

He felt depressed; a hard night lay before him, marches and countermarches, advances won and lost. He did not understand the pilot's tactics; a little further on and they would hit against that blackness, like a wall.

On the rim of the horizon in front he now could see a ghostly flicker, like the glow above a smithy. He tapped Fabien's shoulder, but the pilot did not stir.

Now the first eddies of the distant storm assailed them. The mass of metal heaved gently up, pressing itself against the operator's limbs; and then it seemed to melt away, leaving him for some seconds floating in the darkness, levitated. He clung to the steel bulwarks with both hands. The red lamp in the cockpit was all that remained to him of the world of men and he shuddered to know himself descending helpless into the dark heart of night, with only a little thing, a miner's safety-lamp, to see him through. He dared not disturb the pilot to ask his plans; he tightened his grip on the steel ribs and, bending forward, fixed his eyes upon the pilot's shadowed back.

In that obscurity the pilot's head and shoulders were all that showed themselves. His torso was a block of darkness, inclined a little to the left; his face was set toward the storm, bathed intermittently, no doubt, by flickering gleams. He could not see that face; all the feelings thronging there to meet the onset of the storm were hidden from his eyes; lips set with anger and resolve, a white face holding elemental colloquy with the leaping flashes ahead.

Yet he divined the concentrated force that brooded in that mass of shadow, and he loved it. True, it was carrying him toward the tempest, yet it shielded him. True, those hands, gripping the controls, pressed heavy on the storm, as on some huge beast's neck, but the strong shoulders never budged, attesting vast reserves of force. And after all, he said to himself, the pilot's responsible. So, carried like a pillion-rider on this breakneck gallop into the flames, he could relish to its full the solid permanence, the weight and substance implicit in that dark form before him.

On the left, faint as a far revolving light, a new storm-center kindled.

The wireless operator made as if to touch Fabien's shoulder and warn him, but then he saw him slowly turn his head, fix his eyes a while on this new enemy and then as slowly return to his previous position, his neck pressed back against the leather pad, shoulders unmoving as before.

## VIII

Rivière went out for a short walk, hoping to shake off his malaise, which had returned. He who had only lived for action, dramatic action, now felt a curious shifting of the crisis of the drama, toward his own personality. It came to him that the little people of these little towns, strolling around their bandstands, might seem to lead a placid life and yet it had its tragedies; illness, love, bereavements, and that perhaps— His own trouble was teaching him many things, "opening windows," as he put it to himself.

Toward eleven he was breathing more easily and turned back toward the offices, slowly shouldering his way through the stagnant crowds around the cinemas. He glanced up at the stars which glinted on the narrow street, well-nigh submerged by glaring sky-signs, and said to himself: "To-night, with my two air-mails on their way, I am responsible for all the sky. That star up there is a sign that is looking for me amongst this crowd—and finds me. That's why I'm feeling out of things, a man apart."

A phrase of music came back to him, some notes from a sonata which he had heard the day before in the company of friends. They had not understood. "That stuff bores us and bores you too, only you won't admit it!"

"Perhaps," he had replied.

Then, as to-night, he had felt lonely, but soon had learnt the bounty of such loneliness. The music had breathed to him its message, to him alone amongst these ordinary folk, whispered its gentle secret. And now the star. Across the shoulders of these people a voice was speaking to him in a tongue that he alone could understand.

On the pavement they were hustling him about, "No," he said to himself, "I won't get annoyed. I am like the father of a sick child walking in the crowd, taking short steps, who carries in his breast the hushed silence of his house."

He looked upon the people, seeking to discover which of them, moving with little steps, bore in his heart discovery or love—and he remembered the lighthouse-keeper's isolation.

Back in the office, the silence pleased him. As he slowly walked from one room to another, his footsteps echoed emptiness. The typewriters slept beneath their covers. The big cupboard doors were closed upon the serried files. Ten years of work and effort. He felt as if he were visiting the cellars of a bank where wealth lies heavy on the earth. But these registers contained a finer stuff than gold—a stock of living energy, living but, like the hoarded gold of banks, asleep.

Somewhere he would find the solitary clerk on night duty. Somewhere here a man was working that life and energy should persevere and thus the work goes on from post to post that, from Toulouse to Buenos Aires, the chain of flights should stay unbroken.

"That fellow," thought Rivière, "doesn't know his greatness."

Somewhere, too, the planes were fighting forward; the night flights went on and on like a persistent malady, and on them watch must be kept. Help must be given to these men who with hands and knees and breast to breast were wrestling with the darkness, who knew and only knew an unseen world of shifting things, whence they must struggle out, as from an ocean. And the things they said about it afterwards were—terrible! "I turned the light on to my hands so as to see them." Velvet of hands bathed in a dim red dark-room glow; last fragment, that must be saved, of a lost world.

Rivière opened the door of the Traffic Office. A solitary lamp shone in one corner, making a little pool of light. The clicking of a single typewriter gave meaning to the silence, but did not fill it. Sometimes the telephone buzzed faintly and the clerk on duty rose obedient to its sad, reiterated call. As he took down the receiver that invisible distress was soothed and a gentle, very gentle murmur of voices filled the coign of shadow.

Impassive the man returned to his desk, for drowsiness and solitude had sealed his features on a secret unconfessed. And yet—what menace it may hold, a call from the outer darkness when two postal planes are on their way! Rivière thought of telegrams that invaded the peace of families sitting round their lamp at night and that grief which, for seconds that seem unending, keeps its secret on the father's face. Waves, so weak at first, so distant from the call they carry, and so calm; and yet each quiet purring of the bell held, for Rivière, a faint echo of that cry. Each time the man came back from the shadow toward his lamp, like a diver returning to the surface, the solitude made his movements heavy with their secret, slow as a swimmer's in the undertow.

"Wait! I'll answer."

Rivière unhooked the receiver and a world of murmurs hummed in his ears.

"Rivière speaking."

Confused sounds, then a voice: "I'll put you on the radio station."

A rattle of plugs into the standard, then another voice: "Radio Station speaking. I'll pass you the messages."

Rivière noted them, nodding. "Good. . . . Good . . ."

Nothing important, the usual routine news. Rio de Janeiro asking for information, Montevideo reporting on the weather, Mendoza on the plant. Familiar sounds.

"And the planes?" he asked.

"The weather's stormy. We don't hear them to-night."

"Right!"

The night is fine here and starry, Rivière thought, yet those fellows can detect in it the breath of the distant storm.

"That's all for the present," he said.

As Rivière rose the clerk accosted him: "Papers to sign, sir."

Rivière discovered that he greatly liked this subordinate of his who was bearing, too, the brunt of night. "A comrade in arms," he thought. "But he will never guess, I fancy, how to-night's vigil brings us near each other."

# IX

As he was returning to his private office, a sheaf of papers in his hand, Rivière felt the stab of pain in his right side which had been worrying him for some weeks past.

"That's bad. . . ."

He leaned against the wall a moment.

"It's absurd!"

Then he made his way to his chair.

Once again he felt like some old lion fallen in a trap and a great sadness came upon him.

"To think I've come to this after all those years of work! I'm fifty; all that time I've filled my life with work, trained myself, fought my way, altered the course of events and here's this damned thing getting a hold of me, obsessing me till it seems the only thing that matters in the world. It's absurd!"

He wiped away a drop or two of sweat, waited till the pain had ebbed and settled down to work, examining the memoranda on his table.

"In taking down Motor 301 at Buenos Aires we discovered that . . . The employee responsible will be severely punished."

He signed his name.

"The Florianopolis staff, having failed to comply with orders . . ."

He signed.

"As a disciplinary measure Airport Supervisor Richard, is transferred on the following grounds. . . ."

He signed.

Then, as the pain in his side, slumbering but persistent, new as a new meaning in life, drove his thoughts inward toward himself, an almost bitter mood came over him.

"Am I just or unjust? I've no idea. All I know is that when I hit hard there are fewer accidents. It isn't the individual that's responsible but a sort of hidden force and I can't get at it without—getting at every one! If I were merely just, every night flight would mean a risk of death."

A sort of disgust came over him, that he had given himself so hard a road to follow. Pity is a fine thing, he thought. Lost in his musings, he turned the pages over.

"Roblet, as from this day, is struck off the strength. . . ."

He remembered the old fellow and their talk the evening before.

"There's no way out of it, an example must be made."

"But, sir. . . . It was the only time, just once in a way, sir . . . and I've been hard at it all my life!"

"An example must be made."

"But . . . but, sir. Please see here, sir."

A tattered pocket-book, a newspaper picture showing young Roblet standing beside an aëroplane. Rivière saw how the old hands were trembling upon this little scrap of fame.

"It was in nineteen ten, sir. That was the first plane in Argentina and I assembled it. I've been in aviation since nineteen ten, think of it, sir! Twenty years! So how can you say . . . ? And the young 'uns, sir, won't they just laugh about it in the shop! Won't they just chuckle!"

"I can't help that."

"And my kids, sir. I've a family."

"I told you you could have a job as a fitter."

"But there's my good name, sir, my name . . . after twenty years' experience. An old employee like me!"

"As a fitter."

"No, sir, I can't see my way to that. I somehow can't, sir!"

The old hands trembled and Rivière averted his eyes from their plump, creased flesh which had a beauty of its own.

"No, sir, no. . . . And there's something more I'd like to say."

"That will do."

Not he, thought Rivière, it wasn't he whom I dismissed so brutally, but the mischief for which, perhaps, he was not responsible, though it came to pass through him. For, he mused, we can command events and they obey us; and thus we are creators. These humble men, too, are things and we create them. Or cast them aside when mischief comes about through them.

"There's something more I'd like to say." What did the poor old fellow want to say? That I was robbing him of all that made life dear? That he loved the clang of tools upon the steel of airplanes, that all the ardent poetry of life would now be lost to him . . . and then, a man must live?

"I am very tired," Rivière murmured and his fever rose, insidiously caressing him. "I liked that old chap's face." He tapped the sheet of paper with his finger. It came back to him, the look of the old man's hands and he now seemed to see them shape a faltering gesture of thankfulness. "That's all right," was all he had to say. "That's right. Stay!" And then— He pictured the torrent of joy that would flow through those old hands. Nothing in all the world, it seemed to him, could be more beautiful than that joy revealed not on a face, but in those toil-worn hands. Shall I tear up this paper? He imagined the old man's homecoming to his family, his modest pride.

"So they're keeping you on?"

"What do you think? It was I who assembled the first plane in Argentina!"

The old fellow would get back his prestige, the youngsters cease to laugh.

As he was asking himself if he would tear it up, the telephone rang.

There was a long pause, full of the resonance and depth that wind and distance give to voices.

"Landing-ground speaking. Who is there?"

"Rivière."

"No. 650 is on the tarmac, sir."

"Good."

"We've managed to fix it up, but the electric circuit needed overhauling at the last minute, the connections had been bungled."

"Yes. Who did the wiring?"

"We will inquire and, if you agree, we'll make an example. It's a serious matter when the lights give out on board."

"You're right."

If, Rivière was thinking, one doesn't uproot the mischief whenever and wherever it crops up, the lights may fail and it would be criminal to let it pass when, by some chance, it happens to unmask its instrument; Roblet shall go.

The clerk, who had noticed nothing, was busy with his typewriter.

"What's that?"

"The fortnightly accounts."

"Why not ready?"

"I . . . I . . ."

"We'll see about that."

Curious, mused Rivière, how things take the upper hand, how a vast dark force, the force that thrusts up virgin forests, shows itself whenever a great work is in the making! And he thought of temples dragged asunder by frail liana tendrils.

A great work. . . .

And, heartening himself, he let his thought flow on. These men of mine, I love them; it's not they whom I'm against, but what comes about through them. . . . His heart was throbbing rapidly and it hurt him. . . . No, I cannot say if I am doing right or what precise value should be set on a human life, or suffering, or justice. How should I know the value of a man's joys? Or of a trembling hand? Of kindness, or pity?

Life is so full of contradictions; a man muddles through it as best he can. But to endure, to create, to barter this vile body. . . .

As if to conclude his musings he pressed the bell-push.

"Ring up the pilot of the Europe mail and tell him to come and see me before he leaves."

For he was thinking: I must make sure he doesn't turn back needlessly. If I don't stir my men up the night is sure to make them nervous.

## X

Roused by the call, the pilot's wife looked musingly at her husband. I'll let him sleep a bit longer, she thought.

She admired that spanned bared chest of his and the thought came to her of a well-built ship. In the quiet bed, as in a harbor, he was sleeping and, lest anything should spoil his rest, she smoothed out a fold of the sheet, a little wave of shadow, with her hand, bringing calm upon the bed, as a divine hand calms the sea.

Rising, she opened the window and felt the wind on her face. Their room overlooked Buenos Aires. A dance was going on in a house near by and the music came to her upon the wind, for this was the hour of leisure and amusement. In a hundred thousand barracks this city billeted its men and all was peaceful and secure; but, the woman thought, soon there'll be a cry "To arms!" and only one man—mine—will answer it. True, he rested still, yet his was the ominous rest of reserves soon to be summoned to the front. This town at rest did not protect him; its light would seem as nothing when, like a young god, he rose above its golden dust. She looked at the strong arms which, in an hour, would decide the fortune of the Europe mail, bearing a high responsibility, like a city's fate. The thought troubled her. That this man alone, amongst those millions, was destined for the sacrifice made her sad. It estranged him from her love. She had cherished him, watched over him, caressed him, not for herself but for this night which was to take him. For struggles, fears, and victories which she would never know. Wild things they were, those hands of his, and only tamed to tenderness; their real task was dark to her. She knew this man's smile, his gentle ways of love, but not his godlike fury in the storm. She might snare him in a fragile net of music, love and flowers, but, at each departure, he would break forth without, it seemed to her, the least regret.

He opened his eyes. "What time is it?"

"Midnight."

"How's the weather?"

"I don't know."

He rose and, stretching himself, walked to the window. "Won't be too cold. What's the wind?"

"How should I know?"

He leaned out. "Southerly. That's tophole. It'll hold as far as Brazil anyhow."

He looked at the moon and reckoned up his riches and then his gaze fell upon the town below. Not warm or kind or bright it seemed to him; already in his mind's eye its worthless, shining sands were running out.

"What are you thinking about?"

He was thinking of the fog he might encounter toward Porto Allegre.

"I've made my plans. I know exactly where to turn."

He still was bending down, inhaling deeply like a man about to plunge, naked, into the sea.

"You don't even seem to mind it! How long will you be away?" she asked.

A week or ten days, he couldn't say. "Mind it?" Why should he? All those cities, plains, and mountains. . . . In freedom he was going out to conquer them. In under an hour, he thought, he would have annexed Buenos Aires and tossed it aside!

He smiled at his thoughts. This town . . . it will soon be left behind. It's fine starting out at night. One opens out the gas, facing south, and ten seconds later swings the landscape roundabout, heading up north. The town looks like the bottom of the sea.

She thought of all a man must lay aside to conquer. "So you don't like your home?"

"I do like my home."

But his wife knew that he was already on his way and even now his sturdy shoulders were pressing up against the sky.

She pointed to the sky. "A fine night. See, your road is paved with stars!"

He laughed. "Yes."

She rested her hand on his shoulder and its moist warmth disquieted her; did some danger theaten this young flesh of his?

"I know how strong you are, but—do take care!"

"Of course I'll take care."

Then he began dressing. For the occasion he chose the coarsest, roughest fabrics, the heaviest of leather—a peasant's kit. The heavier he grew, the more she admired him. Herself she buckled his belt, helped to pull his boots on.

"These boots pinch me!"

"Here are the others."

"Bring a cord for my emergency-lamp."

She looked at him, set to rights the last flaw in his armor; all fell into place.

"You look splendid."

Then she noticed that he was carefully brushing his hair.

"For the benefit of the stars?" she questioned.

"I don't want to feel old."

"I'm jealous."

He laughed again and kissed her, pressing her to his heavy garments. Then he lifted her from the ground between his outstretched arms, like a little girl, and, laughing still, deposited her on the bed.

"Go to sleep!"

He shut the door behind him and, passing amongst the indistinguishable folk of night, took the first step toward his conquests.

She remained, sadly looking at these flowers and books, little friendly things which meant for him no more than the bottom of the sea.

## XI

Rivière greeted him.

"That's a nice trick you played on me, your last trip! You turned back though the weather reports were good. You could have pushed through all right. Got the wind up?"

Surprised, the pilot found no answer. He slowly rubbed his hands one on the other. Then, raising his head, he looked Rivière in the eyes.

"Yes," he answered.

Deep in himself Rivière felt sorry for this brave fellow who had been afraid. The pilot tried to explain.

"I couldn't see a thing. No doubt, further on . . . perhaps . . . the radio said. . . . But my lamp was getting weak and I couldn't see my hands. I tried turning on my flying-light so as to spot a wing anyhow, but I saw nothing. It was like being at the bottom of a huge pit, and no getting out of it. Then my engine started a rattle."

"No."

"No?"

"No, we had a look at it. In perfect order. But a man always thinks the engine's rattling when he gets the wind up."

"And who wouldn't? The mountains were above me. When I tried to climb I got caught in heavy squalls. When one can't see a damned thing, squalls, you know. . . . Instead of climbing I lost three hundred feet or more. I couldn't even see the gyroscope or the manometers. It struck me that the engine was running badly and heating up, and the oil-pressure was going down. And it was dark as a plague of Egypt. Damned glad I was to see the lights of a town again."

"You've too much imagination. That's what it is."

The pilot left him.

Rivière sank back into the arm-chair and ran his fingers through his grizzled hair.

The pluckiest of my men, he thought. It was a fine thing he did that night, but I've stopped him from being afraid.

He felt a mood of weakness coming over him again.

To make oneself beloved one need only show pity. I show little pity, or I hide it. Sure enough it would be fine to create friendships and human kindness around me. A doctor can enjoy that in the course of his profession. But I'm the servant of events and, to make others serve them too, I've got to temper my men like steel. That dark necessity is with me every night when I read over the flight reports. If I am slack and let events take charge, trusting to routine, always mysteriously something seems to happen. It is as if my will alone forbade the plane in flight from breaking or the storm to hold the mail up. My power sometimes amazes me.

His thoughts flowed on.

Simple enough, perhaps. Like a gardener's endless labor on his lawn; the mere pressure of his hand drives back into the soil the virgin forest which the earth will engender time and time again.

His thoughts turned to the pilot.

I am saving him from fear. I was not attacking *him* but, across him, that stubborn inertia which paralyzes men who face the unknown. If I listen and sympathize, if I take his adventure seriously, he will fancy he is returning from a land of mystery, and mystery alone is at the root of fear. We must do away with mystery. Men who have gone down into the pit of darkness must come up and say—there's nothing in it! This man must enter the inmost heart of night, that clotted darkness, without even his little miner's davy, whose light, falling only on a hand or wing, suffices to push the unknown a shoulder's breath away.

Yet a silent communion, deep within them, united Rivière and his pilots in the battle. All were like shipmates, sharing a common will to victory.

Rivière remembered other battles he had joined to conquer night. In official circles darkness was dreaded as a desert unexplored. The idea of launching a craft at a hundred and fifty miles an hour against the storm and mists and all the solid obstacles night veils in darkness might suit the military arm; you leave on a fine night, drop bombs and return to your starting-point. But regular night-services were doomed to fail. "It's a matter of life and death," said Rivière, "for the lead we gain by day on ships and railways is lost each night."

Disgusted, he had heard them prate of balance-sheets, insurance and, above all, public opinion. "Public opinion!" he exclaimed. "The public does as it's told!" But it was all waste of time, he was saying to himself. There's something far above all that. A living thing forces its way through, makes its own laws to live and nothing can resist it. Rivière had no notion when or how commercial aviation would tackle the problem of night-flying but its inevitable solution must be prepared for.

Those green table-cloths over which he had leaned, his chin propped on his arm, well he remembered them! And his feeling of power as he heard the others' quibbles! Futile these had seemed, doomed from the outset by the force of life. He felt the weight of energy that gathered in him. And I shall win, thought Rivière, for the weight of argument is on my side. That is the natural trend of things. They urged him to propose a utopian scheme, devoid of every risk. "Experience will guide us to the rules," he said. "You cannot make rules precede practical experience."

After a hard year's struggles, Rivière got his way. "His faith saw him through," said some, but others: "No, his tenacity. Why, the fellow's as obstinate as a bear!" But Rivière put his success down to the fact that he had lent his weight to the better cause.

Safety first was the obsession of those early days. Planes were to leave only an hour before dawn, to land only an hour after sunset. When Rivière felt surer of his ground, then and only then did he venture to send his planes into the depth of night. And now, with few to back him, disowned by nearly all, he plowed a lonely furrow.

Rivière rang up to learn the latest messages from the planes in flight.

## XII

Now the Patagonia mail was entering the storm and Fabien abandoned all idea of circumventing it; it was too widespread for that, he reckoned, for the vista of lightning-flashes led far inland, exposing battlement on battlement of clouds. He decided to try passing below it, ready to beat a retreat if things took a bad turn.

He read his altitude, five thousand five hundred feet, and pressed the controls with his palms to bring it down. The engine started thudding violently, setting all the plane aquiver. Fabien corrected the gliding angle approximately, verifying on the map the height of the hills, some sixteen hundred feet. To keep a safety margin he determined to fly at a trifle above two thousand, staking his altitude as a gambler risks his fortune.

An eddy dragged him down, making the plane tremble still more harshly and he felt the threat of unseen avalanches that toppled all about him. He dreamt an instant of retreat and its guerdon of a hundred thousand stars, but did not shift his course by one degree.

Fabien weighed his chances; probably this was just a local storm, as Trelew, the next halt, was signaling a sky only three-quarters overcast. A bare twenty minutes more of solid murk and he would be through with it. Nevertheless the pilot felt uneasy. Leaning to his left, to windward, he sought to catch those vague gleams which, even in darkest nights, flit here and there. But even those vagrant gleams were gone; at most there lingered patches in the mass of shadow where the night seemed less opaque, or was it only that his eyes were growing strained?

The wireless operator handed him a slip of paper.

"Where are we?"

Fabien would have given much to know. "Can't say exactly," he answered. "We are flying by compass across a storm."

He leaned down again. The flame from the exhaust was getting on his nerves. There it was, clinging to the motor like a spray of fire-flowers, so pale it seemed that moonlight would have quelled it, but, in this nothingness, engulfing all the visible world. He watched it streaming stiffly out into the wind, like a torch-flame.

Every thirty seconds Fabien bent down into the cockpit to check the gyroscope and compass. He dared not light the dim red lamps which would have dazzled his eyes for some moments, but the luminous dial-hands were ceaselessly emitting their pale and starry radiance. And in all those needles and printed figures the pilot found an illusive reassurance, as in the cabin of a ship swept by the waves. For, like a very sea of strange fatality, the night was rolling up against him with all its rocks and reefs and wreckage.

"Where are we?" the operator asked again.

Fabien drew himself up and, leaning to the left, resumed his tremendous vigil. He had no notion left how many hours more and what efforts would be needed to deliver him from fettering darkness. Would he ever come clear, he wondered, for he was staking his life on this little slip of dirty, crumpled paper, which he unfolded and re-read a thousand times to nurse his hopes: *Trelew. Sky three-quarters overcast. Westerly breeze.* If there still remained a clear patch over Trelew, he would presently glimpse its lights across a cloud-rift. Unless. . . .

That promise of a faint gleam far ahead beckoned him on; but, to make sure, he scribbled a message to the radio operator. "Don't know if I can get through. Ask if the weather's holding out behind."

The answer appalled him.

"Commodoro reports: Impossible return here. Storm."

He was beginning to measure this unforeseen offensive, launched from the Cordillera toward the sea. Before he could make them the storm would have burst upon the cities.

"Get the San Antonio weather report."

"San Antonio reports: West wind rising. Storm in the west. Sky three-quarters overcast. San Antonio picking up badly on account of interferences. I'm having trouble too. I shall have to pull up the aërial on account of the lightning. Will you turn back? What are your plans?"

"Stow your damned questions! Get Bahia Blanca!"

"Bahia Blanca reports: Violent westerly gale over Bahia Blanca expected in less than twenty minutes."

"Ask Trelew."

"Trelew reports: Westerly gale; a hundred feet per second; rain squalls."

"Inform Buenos Aires: We are cut off on all sides; storm developing over a depth of eight hundred miles; no visibility. What shall we do?"

A shoreless night, the pilot thought, leading to no anchorage (for every port was unattainable, it seemed), nor toward dawn. In an hour and twenty minutes the fuel would run out. Sooner or later he must blindly founder in the sea of darkness. Ah, if only he could have won through to daylight!

Fabien pictured the dawn as a beach of golden sand where a

man might get a foothold after this hard night. Beneath him the plains, like friendly shores, would spread their safety. The quiet land would bear its sleeping farms and flocks and hills. And all the flotsam swirling in the shadows would lose its menace. If it were possible, how gladly he would swim toward the strand of daylight! But, well he knew, he was surrounded; for better or for worse the end would come within this murk of darkness. . . . Sometimes, indeed, when daybreak came, it seemed like convalescence after illness.

What use to turn his eyes towards the east, home of the sun? Between them lay a gulf of night so deep that he could never clamber up again.

## XIII

"The Asuncion mail is making good headway; it should be in at about two. The Patagonia mail, however, seems to be in difficulties and we expect it to be much overdue."

"Very good, Monsieur Rivière."

"Quite possibly we won't make the Europe mail wait for it; as soon as Asuncion's in, come for instructions, please. Hold yourself in readiness."

Rivière read again the weather reports from the northern sectors. "Clear sky; full moon; no wind." The mountains of Brazil were standing stark and clear against the moonlit sky, the tangled tresses of their jet-black forests falling sheer into a silver tracery of sea. Upon those forests the moonbeams played and played in vain, tingeing their blackness with no light. Black, too, as drifting wreckage, the islands flecked the sea. But all the outward air-route was flooded by that exhaustless fountain of moonlight.

If Rivière now gave orders for the start, the crew of the Europe mail would enter a stable world, softly illuminated all night long. A land which held no threat for the just balance of light and shade, unruffled by the least caress of those cool winds which, when they freshen, can ruin a whole sky in an hour or two.

Facing this wide radiance, like a prospector eyeing a forbidden gold-field, Rivièra hesitated. What was happening in the south put Rivièra, sole protagonist of night flights, in the wrong. His opponents would make such moral capital out of a disaster in Patagonia that all Rivière's faith would henceforth be unavailing. Not that his faith wavered; if, through a fissure in his work, a tragedy had entered in, well, the tragedy might prove the fissure—but it proved nothing else. Perhaps, he thought, it would be well to have look-out posts in the west. That must be seen to. "After all," he said to himself, "my previous arguments hold good as ever and the possibilities of accident are reduced by one, the one to-night has illustrated." The strong are strengthened by reverses; the trouble is that the true meaning of events scores next to nothing in the match we play with men. Appearances decide our gains or losses and the points are trumpery. And a mere semblance of defeat may hopelessly checkmate us.

He summoned an employee. "Still no radio from Bahia Blanca?"

"No."

"Ring up the station on the phone."

Five minutes later he made further inquiries. "Why don't you pass on the messages?"

"We can't hear the mail."

"He's not sending anything?"

"Can't say. Too many storms. Even if he was sending we shouldn't pick it up."

"Can you get Trelew?"

"We can't hear Trelew."

"Telephone."

"We've tried. The line's broken."

"How's the weather your end?"

"Threatening. Very sultry. Lightning in the west and south."

"Wind?"

"Moderate so far. But in ten minutes the storm will break; the lightning's coming up fast."

Silence.

"Hullo, Bahia Blanca! You hear me? Good. Call me again in ten minutes."

Rivière looked through the telegrams from the southern stations. All alike reported: No message from the plane. Some had ceased by now to answer Buenos Aires and the patch of silent areas was spreading on the map as the cyclone swept upon the little towns and one by one, behind closed doors, each house along the lightless streets grew isolated from the outer world, lonely as a ship on a dark sea. And only dawn would rescue them.

Rivière, poring on the map, still hoped against hope to discover a haven of clear sky, for he had telegraphed to the police at more than thirty up-country police-stations and their replies were coming in. And the radio-posts over twelve hundred miles of country had orders to advise Buenos Aires within thirty seconds if any message from the plane was picked up, so that Fabien might learn at once whither to fly for refuge.

The employees had been warned to attend at 1 A.M. and were now at their posts. Somehow, mysteriously, a rumor was gaining ground that perhaps the night flights would be suspended in future and the Europe mail would leave by day. They spoke in whispers of Fabien, the cyclone and, above all, of Rivière whom they pictured near at hand and point by point capitulating to this rebuff the elements had dealt.

Their chatter ceased abruptly; Rivière was standing at his door, his overcoat tight-buttoned across his chest, his hat well down upon his eyes, like the incessant traveler he always seemed. Calmly he approached the head clerk.

"It's one ten. Are the papers for the Europe mail in order?"

"I—I thought—"

"Your business is to carry out orders, not to think."

Slowly turning away, he moved toward an open window, his hands clasped behind his back. A clerk came up to him.

"We have very few replies, sir. We hear that a great many telegraph lines in the interior have been destroyed."

"Right!"

Unmoving, Rivière stared out into the night.

Thus each new message boded new peril for the mail. Each town, when a reply could be sent through before the lines were broken, announced the cyclone on its way, like an invading horde. "It's coming up from the Cordillera, sweeping everything before it, toward the sea."

To Rivière the stars seemed over-bright, the air too moist. Strange night indeed! It was rotting away in patches, like the substance of a shining fruit. The stars, in all their host, still looked down on Buenos Aires—an oasis, and not to last. A haven out of Fabien's range, in any case. A night of menace, touched and tainted by an evil wind. A difficult night to conquer.

Somewhere in its depths an airplane was in peril; here, on the margin, they were fighting to rescue it, in vain.

## XIV

Fabien's wife telephoned.

Each night she calculated the progress of the homing Patagonia mail. "He's leaving Trelew now," she murmured. Then went to sleep again. Presently: "He's getting near San Antonio, he has its lights in view." Then she got out of bed, drew back the curtains and summed up the sky. "All those clouds will worry him." Sometimes the moon was wandering like a shepherd and the young wife was heartened by the faithful moon and stars, the thousand presences that watched her husband. Toward one o'clock she felt him near her. "Not far to go, Buenos Aires is in sight." Then she got up again, prepared a meal for him, a nice steaming cup of coffee. "It's so cold up there!" She always welcomed him as if he had just descended from a snow-peak. "You *must* be cold!" "Not a bit." "Well, warm yourself anyhow!" She had everything ready at a quarter past one. Then she telephoned. To-night she asked the usual question.

"Has Fabien landed?"

The clerk at the other end grew flustered. "Who's speaking?"

"Simone Fabien."

"Ah! A moment, please. . . ."

Afraid to answer, he passed the receiver to the head clerk.

"Who's that?"

"Simone Fabien."

"Yes. What can I do for you?"

"Has my husband arrived?"

After a silence which must have baffled her, there came a monosyllable. "No."

"Is he delayed?"

"Yes."

Another silence. "Yes, he is delayed."

"Ah!"

The cry of a wounded creature. A little delay, that's nothing much, but when it lasts, when it lasts. . . .

"Yes. And when—when is he expected in?"

"When is he expected? We . . . we don't know exactly . . ."

A solid wall in front of her, a wall of silence, which only gave her back the echo of her questions.

"Do please tell me, where is he now?"

"Where is he? Wait. . . ."

This suspense was like a torture. Something was happening there, behind that wall.

At last, a voice! "He left Commodoro at seven thirty this evening."

"Yes? And then?"

"Then—delayed, seriously delayed by stormy weather."

"Ah! A storm!"

The injustice of it, the sly cruelty of that moon up there, that lazing moon of Buenos Aires! Suddenly she remembered that it took barely two hours to fly from Commodoro to Trelew.

"He's been six hours on the way to Trelew! But surely you've had messages from him. What does he say?"

"What does he say? Well, you see, with weather like that . . . it's only natural . . . we can't hear him."

"Weather like—?"

"You may rest assured, madame, the moment we get news of him, we will ring you up."

"Ah! You've no news."

"Good-night, madame."

"No! No! I want to talk to the director."

"I'm sorry, he's very busy just now; he has a meeting on—"

"I can't help that. That doesn't matter. I insist on speaking to him."

The head clerk mopped his forehead. "A moment, please."

He opened Rivière's door.

"Madame Fabien wants to speak to you, sir."

"Here," thought Rivière, "is what I was dreading." The emotional elements of the drama were coming into action. His first impulse was to thrust them aside; mothers and women are not allowed in an operating theater. And all emotion is bidden to hold its peace on a ship in peril; it does not help to save the crew. Nevertheless he yielded.

"Switch on to my phone."

No sooner did he hear that far off, quavering voice, than he knew his inability to answer it. It would be futile for both alike, worse than futile, to meet each other.

"Do not be alarmed, madame, I beg you. In our calling it so often happens that a long while passes without news."

He had reached a point where not the problem of a small personal grief but the very will to act was in itself an issue. Not so much Fabien's wife as another theory of life confronted Rivière now. Hearing that timid voice, he could but pity its infinite distress—and know it for an enemy! For action and individual happiness have no truck with each other; they are eternally at war. This woman, too, was championing a self-coherent world with its own rights and duties, that world where a lamp shines at nightfall on the table, flesh calls to mated flesh, a homely world of love and hopes and memories. She stood up for her happiness and she was right. And Rivière, too, was right, yet he found no words to set against this woman's truth. He was discovering the truth within him, his own inhuman and unutterable truth, by an humble light, the lamplight of a little home!

"Madame . . . !"

She did not hear him. Her hands were bruised with beating on the wall and she lay fallen, or so it seemed to him, almost at his feet.

One day an engineer had remarked to Rivière, as they were bending above a wounded man, beside a bridge that was being erected: "Is the bridge worth a man's crushed face?" Not one of the peasants using the road would ever have wished to mutilate this face so hideously just to save the extra walk to the next bridge. "The welfare of the community," the engineer had continued, "is just the sum of individual welfares and has no right to look beyond them." "And yet," Rivière observed on a subsequent occasion, "even though human life may be the most precious thing on earth, we always behave as if there were something of higher value than human life. . . . But what thing?"

Thinking of the lost airmen, Rivière felt his heart sink. All man's activity, even the building of a bridge, involves a toll of suffering and he could no longer evade the issue—"Under what authority?"

These men, he mused, who perhaps are lost, might have led happy lives. He seemed to see as in a golden sanctuary the evening lamplight shine on faces bending side by side. "Under what authority have I taken them from all this?" he wondered. What was his right to rob them of their personal happiness? Did not the highest of all laws ordain that these human joys should be safeguarded? But he destroyed them. And yet one day, inevitably, those golden sanctuaries vanish like mirage. Old age and death, more pitiless than even he, destroy them. There is, perhaps, some other thing, something more lasting, to be saved; and, perhaps, it was to save this part of man that Rivière was working. Otherwise there could be no defense for action.

To love, only to love, leads nowhere. Rivière knew a dark sense of duty, greater than that of love. And deep within it there might lie another emotion and a tender one, but worlds away from ordinary feelings. He recalled a phrase that he once had read: "The one thing is to make them everlasting. . . . That which you seek within yourself will die." He remembered a temple of the sun-god, built by the ancient Incas of Peru. Tall menhirs on a mountain. But for these what would be left of all that mighty civilization which with its massive stones weighs heavy, like a dark regret, on modern man? Under the mandate of what strange love, what ruthlessness, did that primeval leader of men compel his hordes to drag this temple up the mountainside, bidding them raise up their eternity? And now another picture rose in Rivière's mind; the people of the little towns, strolling by nights around their bandstands. That form of happiness, those shackles . . . he thought. The leader of those ancient races may have had scant compassion for man's sufferings, but he had a boundless pity for his death. Not for his personal death, but pity for his race, doomed to be blotted out beneath a sea of sand. And so he bade his folk set up these stones at least, something the desert never would engulf.

## XV

That scrap of folded paper might perhaps save him yet; gritting his teeth, Fabien unfolded it.

"Impossible communicate Buenos Aires. Can't even touch the key, the shocks are numbing my hands."

In his vexation Fabien wanted to reply, but the moment his hands left the controls to write, a vast groundswell seemed to surge up across his body; the eddies lifted him in his five tons of metal and rocked him to and fro. He abandoned the attempt.

Again he clenched his hands upon the tempest and brought it down. Fabien was breathing heavily. If that fellow pulled up the aërial for fear of the storm, Fabien would smash his face in when they landed. At all costs they must get in touch with Buenos Aires—as though across the thousand miles and more a safety-line might be flung to rescue them from this abyss! If he could not have one vagrant ray of light, not even the flicker of an inn-lamp—of little help indeed, yet shining like a beacon, earnest of the earth—at least let him be given a voice, a single word from that lost world of his. The pilot raised his fist and shook it in the red glow, hoping to make the man behind him understand the tragic truth, but the other was bending down to watch a world in ruins, with its buried cities and dead lights, and did not see him.

Let them shout any order whatever to him and Fabien would obey. If they tell me to go round and round, he thought, I'll turn in circles and if they say I must head due south. . . . For somewhere, even now, there still were lands of calm, at peace beneath the wide moon-shadows. His comrades down there, omniscient folk like clever scientists, knew all about them, poring upon the maps beneath their hanging lamps, pretty as flower-bells. But he, what could he know save squalls and night, this night that buffeted him with its swirling spate of darkness? Surely they could not leave two men to their fate in these whirlwinds and flaming clouds! No, that was unthinkable! They might order Fabien to set his course at two hundred and forty degrees, and he would do it. . . . But he was alone.

It was as if dead matter were infected by his exasperation; at every plunge the engine set up such furious vibrations that all the fuselage seemed convulsed with rage. Fabien strained all his efforts to control it; crouching in the cockpit, he kept his eyes fixed on the artificial horizon only, for the masses of sky and land outside were not to be distinguished, lost both alike in a welter as of worlds in the making. But the hands of the flying instruments oscillated more and more abruptly, grew almost impossible to follow. Already the pilot, misled by their vagaries, was losing altitude, fighting against odds, while deadly quicksands sucked him down into the darkness. He read his height, sixteen hundred—just the level of the hills. He guessed their towering billows hard upon him, for now it seemed that all these earthen monsters, the least of which could crush him into nothingness, were breaking loose from their foundations and careering about in a drunken frenzy. A dark tellurian carnival was thronging close and closer round him.

He made up his mind. He would land no matter where, even if it meant cracking up! To avoid the hills anyhow, he launched his only landing flare. It sputtered and spun, illumining a vast plain, then died away; beneath him lay the sea!

His thoughts came quickly. Lost—forty degrees' drift—yes, I've drifted, sure enough—it's a cyclone—where's land? He turned due west. Without another flare, he thought, I'm a goner. Well, it was bound to happen one day. And that fellow behind there! Sure thing he's pulled up the aërial. . . . But now the pilot's anger had ebbed away. He had only to unclasp his hands and their lives would slither through his fingers like a trivial mote of dust. He held the beating heart of each—his own, his comrade's—in his hands. And suddenly his hands appalled him.

In these squalls that battered on the plane, to counteract the jerks of the wheel, which else would have snapped the control cables, he clung to it with might and main, never relaxing his hold for an instant. But now he could no longer feel his hands, numbed by the strain. He tried to shift his fingers and get some signal they were there, but he could not tell if they obeyed his will. His arms seemed to end in two queer foreign bodies, insentient like flabby rubber pads. "Better try hard to think I'm gripping," he said to himself. But whether his thought carried as far as his hands he could not guess. The tugs upon the wheel were only felt by him as sudden twinges in his shoulders. "I'll let go for sure. My fingers will open." His rashness scared him—that he had dared to even think such words!—for now he fancied that his hands, yielding to the dark suggestion of his thought, were opening slowly, slowly opening in the shadow, to betray him.

He might keep up the struggle, chance his luck; no destiny attacks us from outside. But, within him, man bears his fate and there comes a moment when he knows himself vulnerable; and then, as in a vertigo, blunder upon blunder lures him.

And, at this very moment, there gleamed above his head, across a storm-rift, like a fatal lure within a deep abyss, a star or two.

Only too well he knew them for a trap. A man sees a few stars at the issue of a pit and climbs toward them, and then—never can he get down again but stays up there eternally, chewing the stars. . . .

But such was his lust for light that he began to climb.

## XVI

He climbed and it grew easier to correct the plunges for the stars gave him his bearings. Their pale magnet drew him up; after that long and bitter quest for light, for nothing in the world would he forego the frailest gleam. If the glimmer of a little inn were all his riches, he would turn around this token of his heart's desire until his death! So now he soared toward the fields of light.

Little by little he spiraled up, out of the dark pit which closed again beneath him. As he rose the clouds began to shed their slime of shadow, flowing past him in cleaner, whiter billows. Fabien rose clear.

And now a wonder seized him; dazzled by that brightness, he had to keep his eyes closed for some seconds. He had never dreamt the night-clouds could dazzle thus. But the full moon and all the constellations were changing them to waves of light.

In a flash, the very instant he had risen clear, the pilot found a peace that passed his understanding. Not a ripple tilted the plane but, like a ship that has crossed the bar, it moved across a tranquil anchorage. In an unknown and secret corner of the sky it floated, as in a harbor of the Happy Isles. Below him still the storm was fashioning another world, thridded with squalls and cloudbursts and lightnings, but turning to the stars a face of crystal snow.

Now all grew luminous, his hands, his clothes, the wings, and Fabien thought that he was in a limbo of strange magic; for the light did not come down from the stars but welled up from below, from all that snowy whiteness.

The clouds beneath threw up the flakes the moon was pouring on them; on every hand they loomed like towers of snow. A milky stream of light flowed everywhere, laving the plane and crew. When Fabien turned he saw the wireless operator smile.

"That's better!" he cried.

But his words were drowned by the rumor of the flight; they conversed in smiles. I'm daft, thought Fabien, to be smiling, we're lost.

And yet—at last a myriad dark arms had let him go; those

bonds of his were loosed, as of a prisoner whom they let walk a while in liberty amongst the flowers.

"Too beautiful," he thought. Amid the far-flung treasure of the stars he roved, in a world where no life was, no faintest breath of life, save his and his companion's. Like plunderers of fabled cities they seemed, immured in treasure-vaults whence there is no escape. Amongst these frozen jewels they were wandering, rich beyond all dreams, but doomed.

# XVII

One of the wireless operators at the Commodoro Rivadavia station in Patagonia made a startled gesture and all the others keeping helpless vigil there crowded round to read the message.

A harsh light fell upon the blank sheet of paper over which they bent. The operator's hand seemed loath to do its task and his pencil shook. The words to write were prisoned in his hand, but already his fingers twitched.

"Storms?"

He nodded assent; he could hardly hear for interferences. Then he scrawled some illegible signs, then words; then, at last, the text came out.

"Cut off at 12,000 feet, above the storm. Proceeding due west toward interior; found we had been carried above sea. No visibility below. Impossible know if still flying over sea. Report if storm extends interior."

By reason of the storms the telegram had to be relayed from post to post to Buenos Aires, bearing its message through the night like bale-fires lit from tower to tower.

Buenos Aires transmitted a reply. "Storm covers all interior area. How much gasoline left?"

"For thirty minutes." These words sped back from post to post to Buenos Aires.

In under half an hour the plane was doomed to plunge into a cyclone which would crash it to the earth.

## XVIII

Rivière was musing, all hope lost; somewhere this plane would founder in the darkness. A picture rose in his mind of a scene which had impressed him in his boyhood; a pond that was being emptied to find a body. Thus, till this flood of darkness had been drained off the earth and daylight turned toward the plains and cornfields, nothing would be found. Then some humble peasants perhaps would come on two young bodies, their elbows folded on their faces, like children asleep amid the grass and gold of some calm scene. Drowned by the night.

Rivière thought of all the treasure buried in the depths of night, as in deep, legendary seas. Night's apple-trees that wait upon the dawn with all their flowers that serve as yet no purpose. Night, perfume-laden, that hides the lambs asleep and flowers that have no color yet.

Little by little the lush tilth, wet woods, and dew-cool meadows would swing toward the light. But somewhere in the hills, no longer dark with menace, amid the fields and flocks, a world at peace again, two children would seem to sleep. And something would have flowed out of the seen world into that other.

Rivière knew all the tenderness of Fabien's wife, the fears that haunted her; this love seemed only lent her for a while, like a toy to some poor child. He thought of Fabien's hand which, firm on the controls, would hold the balance of his fate some minutes yet; that hand had given caresses and lingered on a breast, wakening a tumult there; a hand of godlike virtue, it had touched a face, transfiguring it. A hand that brought miracles to pass.

Fabien was drifting now in the vast splendor of a sea of clouds, but under him there lay eternity. Among the constellations still he had his being, their only denizen. For yet a while he held the universe in his hand, weighed it at his breast. That wheel he clutched upbore a load of human treasure and desperately, from one star to the other, he trafficked this useless wealth, soon to be his no more.

A single radio post still heard him. The only link between him and the world was a wave of music, a minor modulation. Not a lament, no cry, yet purest of sounds that ever spoke despair.

## XIX

Robineau broke in upon his thoughts.

"I've been thinking, sir. . . . Perhaps we might try—"

He had nothing really to suggest but thus proclaimed his good intentions. A solution, how he would have rejoiced to find it! He went about it as if it were a puzzle to be solved. Solutions were his *forte,* but Rivière would not hear of them. "I tell you, Robineau, in life there are no solutions. There are only motive forces, and our task is to set them acting—then the solutions follow." The only force that Robineau had to activate was one which functioned in the mechanics' shop; a humble force which saved propeller-bosses from rusting.

But this night's happenings found Robineau at fault. His inspectorial mandate could not control the elements, nor yet a phantom ship that, as things were, struggled no longer to win a punctuality-bonus but only to evade a penalty which canceled all that Robineau imposed, the penalty of death.

There was no use for Robineau now and he roamed the offices, forlorn.

Rivière was informed that Fabien's wife wished to see him. Tormented by anxiety, she was waiting in the clerks' office till Rivière could receive her. The employees were stealing glances at her face. She felt shy, almost shamefast, and gazed nervously around her; she had no right of presence here. They went about their tasks as usual and to her it was as if they were trampling on a corpse; in their ledgers no human sorrow but dwindled to dross of brittle figures. She looked for something that might speak to her of Fabien; at home all things confessed his absence—the sheets turned back upon the bed, the coffee on the table, a vase of flowers. Here there was nothing of him; all was at war with pity, friendship, memories. The only word she caught (for in her presence they instinctively lowered their voices) was the oath of an employee clamoring for an invoice. "The dynamo account, God blast you! The one we send to Santos." Raising her eyes she gazed toward this man with a look of infinite wonder. Then to the wall where a map hung. Her lips trembled a little, almost imperceptibly.

The realization irked her that in this room she was the envoy of a hostile creed and almost she regretted having come; she would have liked to hide somewhere and, fearful of being remarked, dared neither cough nor weep. She felt her presence here misplaced, indecent, as though she were standing naked before them. But so potent was *her* truth, the truth within her, that furtively their eyes strayed ever and again in her direction, trying to read it on her face. Beauty was hers and she stood for a holy thing, the world of human happiness. She vouched for the sanctity of that material something with which man tampers when he acts. She closed her eyes before their crowded scrutiny, revealing all the peace which in his blindness man is apt to shatter.

Rivière admitted her.

So now she was come to make a timid plea for her flowers, the coffee waiting on the table, her own young body. Again, in this room, colder even than the others, her lips began to quiver. Thus, too, she bore witness to her truth, unutterable in this alien world. All the wild yearning of her love, her heart's devotion, seemed here invested with a selfish, pestering aspect. And again she would have liked to leave this place.

"I am disturbing you—"

"No," said Rivière, "you are not disturbing me. But unfortunately neither you nor I can do anything except—wait."

There was a faint movement of her shoulders and Rivière guessed its meaning. "What is the use of that lamp, the dinner waiting, and the flowers there when I return?" Once a young mother had confided in Rivière. "I've hardly realized my baby's death as yet. It's the little things that are so cruel—when I see the baby-clothes I had ready, when I wake up at night and there rises in my heart a tide of love, useless now, like my milk . . . all useless!" And for this woman here, Fabien's death would only just begin to-morrow—in every action, useless now, in trivial objects . . . useless. Little by little Fabien would leave his home. A deep, unuttered pity stirred in Rivière's heart.

"Madame—"

The young wife turned and left him with a weak smile, an almost humble smile, ignoring her own power.

Rivière sat down again rather heavily. "Still she is helping me to discover the thing I'm looking for."

He fingered absent-mindedly the messages from the northern airports. "We do not pray for immortality," he thought, "but only not to see our acts and all things stripped suddenly of all their meaning; for then it is the utter emptiness of everything reveals itself."

His gaze fell on the telegrams.

"These are the paths death takes to enter here—messages that have lost their meaning."

He looked at Robineau. Meaningless, too, this fellow who served no purpose now. Rivière addressed him almost gruffly.

"Have I got to tell you what your duties are?"

Then he pushed open the door that led into the Business Office and saw how Fabien's disappearance was recorded there in signs his wife could not have noticed. The slip marked *R.B.903*, Fabien's machine, was already inserted in the wall-index of Unavailable Plant. The clerks preparing the papers for the Europe mail were working slackly, knowing it would be delayed. The airport was ringing up for orders respecting the staff on night duty whose presence was no longer necessary. The functions of life were slowing down. That is death! thought Rivière. His work was like a sailing-ship becalmed upon the sea.

He heard Robineau speaking. "Sir, they had only been married six weeks."

"Get on with your work!"

Rivière, watching the clerks, seemed to see beyond them the workmen, mechanics, pilots, all who had helped him in his task, with the faith of men who build. He thought of those little cities of old time where men had murmured of the "Indies," built a ship and freighted it with hopes. That men might see their hope outspread its wings across the sea. All of them magnified, lifted above themselves and saved—by a ship! He thought: The goal, perhaps, means nothing, it is the thing done that delivers man from death. By their ship those men will live.

Rivière, too, would be fighting against death when he restored to those telegrams their full meaning, to these men on night duty their unrest and to his pilots their tragic purpose; when life itself would make his work alive again, as winds restore to life a sailing-ship upon the sea.

## XX

Commodoro Rivadavia could hear nothing now, but twenty seconds later, six hundred miles away, Bahia Blanca picked up a second message.

"Coming down. Entering the clouds. . . ."

Then two words of a blurred message were caught at Trelew.

". . . see nothing . . ."

Short waves are like that; here they can be caught, elsewhere is silence. Then, for no reason, all is changed. This crew, whose position was unknown, made itself heard by living ears, from somewhere out of space and out of time, and at the radio station phantom hands were tracing a word or two on this white paper.

Had the fuel run out already or was the pilot, before catastrophe, playing his last card: to reach the earth again without a crash?

Buenos Aires transmitted an order to Trelew.

"Ask him."

The radio station looked like a laboratory with its nickel and its copper, manometers and sheaves of wires. The operators on duty in their white overalls seemed to be bending silently above some simple experiment. Delicately they touched their instruments, exploring the magnetic sky, dowsers in quest of hidden gold.

"No answer?"

"No answer."

Perhaps they yet might seize upon its way a sound that told of life. If the plane and its lights were soaring up to join the stars, it might be they would hear a sound—a singing star!

The seconds flowed away, like ebbing blood. Were they still in flight? Each second killed a hope. The stream of time was wearing life away. As for twenty centuries it beats against a temple, seeping through the granite, and spreads the fane in ruin, so centuries of wear and tear were thronging in each second, menacing the airmen.

Every second swept something away; Fabien's voice, his laugh, his smile. Silence was gaining ground. Heavy and heavier silence drowned their voices, like a heavy sea.

"One forty," some one murmured. "They're out of fuel. They can't be flying any more."

Then silence.

A dry and bitter taste rose on their lips, like the dry savor of a journey's end. Something mysterious, a sickening thing, had come to pass. And all the shining nickel and trellised copper seemed tarnished with the gloom that broods on ruined factories. All this apparatus had grown clumsy, futile, out of use; a tangle of dead twigs.

One thing remained; to wait for daybreak. In a few hours all Argentina would swing toward the sun, and here these men were standing, as on a beach, facing the net that was being slowly,

slowly drawn in toward them, none knowing what its take would be.

To Rivière in his office came that quiet aftermath which follows only on great disasters, when destiny has spent its force. He had set the police of the entire country on the alert. He could do no more; only wait.

But even in the house of death order must have its due. Rivière signed to Robineau.

"Circular telegram to the northern airports. *Considerable delay anticipated Patagonia mail. To avoid undue delay Europe mail, will ship Patagonia traffic on following Europe mail.*"

He stooped a little forward. Then, with an effort, he called something to mind, something important. Yes, that was it. Better make sure.

"Robineau!"

"Sir."

"Issue an order, please. Pilots forbidden to exceed 1900 revs. They're ruining my engines."

"Very good, sir."

Rivière bowed his head a little more. To be alone—that was his supreme desire.

"That's all, Robineau. Trot off, old chap!"

And this, their strange equality before the shades, filled Robineau with awe.

## XXI

Robineau was drifting aimlessly about the office. He felt despondent. The company's life had come to a standstill, since the Europe mail, due to start at two, would be countermanded and only leave at daybreak. Morosely the employees kept their posts, but their presence now was purposeless. In steady rhythm the weather reports from the north poured in, but their "no wind," "clear sky," "full moon" evoked the vision of a barren kingdom. A wilderness of stones and moonlight. As Robineau, hardly aware what he was up to, was turning over the pages of a file on which the office superintendent was at work, he suddenly grew conscious that the official in question was at his side, waiting with an air of mocking deference to get his papers back. As if he were saying: "That's my show. Suppose you leave me to it, eh?"

Shocked though he was by his subordinate's demeanor, the inspector found himself tongue-tied and, with a movement of annoyance, handed back the documents. The superintendent resumed his seat with an air of grand punctilio. "I should have told him to go to the devil," thought Robineau. Then, to save his face, he moved away and his thoughts returned to the night's tragedy. For with this tragedy all his chief's campaign went under and Robineau lamented a twofold loss.

The picture of Rivière alone there in his private office rose in Robineau's mind; "old chap," Rivière had said. Never had there been a man so utterly unfriended as he, and Robineau felt an infinite compassion for him. He turned over in his mind vague sentences that hinted sympathy and consolation, and the impulse prompting him struck Robineau as eminently laudable. He knocked gently at the door. There was no answer. Not daring in such a silence to knock louder, he turned the handle. Rivière was there. For the first time Robineau entered Rivière's room almost on an equal footing, almost as a friend; he likened himself to the N.C.O. who joins his wounded general under fire, follows him in defeat and, in exile, plays a brother's part. "Whatever happens I am with you"—that was Robineau's unspoken message.

Rivière said nothing; his head was bowed and he was staring at his hands. Robineau's courage ebbed and he dared not speak; the old lion daunted him, even in defeat. Phrases of loyalty, of ever-growing fervor, rose to his lips; but every time he raised his eyes they encountered that bent head, gray hair and lips tight-set upon their bitter secret. At last he summoned up his courage.

"Sir!"

Rivière raised his head and looked at him. So deep, so far away had been his dream that till now he might well have been unconscious of Robineau's presence there. And what he felt, what was that dream and what his heart's bereavement, none would ever know. . . . For a long while Rivière looked at Robineau as at the living witness of some dark event. Robineau felt ill at ease. An enigmatic irony seemed to shape itself on his chief's lips as he watched Robineau. And the longer his chief watched him, the more deeply Robineau blushed and the more it grew on Rivière that this fellow had come, for all his touching and unhappily sincere good-will, to act as spokesman for the folly of the herd.

Robineau by now had quite lost his bearings. The N.C.O., the general, the bullets—all faded into mist. Something inexplicable was in the air. Rivière's eyes were still intent on him. Reluctantly he shifted his position, withdrew his hand from his pocket. Rivière's eyes were on him still. At last, hardly knowing what he said, he stammered a few words.

"I've come for orders, sir."

Composedly Rivière pulled out his watch. "It is two. The Asuncion mail will land at two ten. See that the Europe mail takes off at two fifteen."

Robineau bruited abroad the astounding news; the night flight would continue. He accosted the office superintendent.

"Bring me that file of yours to check."

The superintendent brought the papers.

"Wait!"

And the superintendent waited.

## XXII

The Asuncion mail signaled that it was about to land. Even at the darkest hour, Rivière had followed, telegram by telegram, its well-ordered progress. In the turmoil of this night he hailed it as the avenger of his faith, an all-conclusive witness. Each message telling of this auspicious flight augured a thousand more such flights to come. "And, after all," thought Rivière, "we don't get a cyclone every night! Once the trail is blazed, it must be followed up."

Coming down, flight by flight, from Paraguay, as from an enchanted garden set with flowers, low houses and slow waters, the pilot had just skirted the edge of a cyclone which never masked from him a single star. Nine passengers, huddled in their traveling-rugs, had pressed their foreheads on the window, as if it were a shop-front glittering with gems. For now the little towns of Argentina were stringing through the night their golden beads, beneath the paler gold of the star-cities. And at his prow the pilot held within his hands his freight of lives, eyes wide open, full of moonlight, like a shepherd. Already Buenos Aires was dyeing the horizon with pink fires, soon to flaunt its diadem of jewels, like some fairy hoard. The wireless operator strummed with nimble fingers the final telegrams, last notes of a sonata he had played *allegro* in the sky—a melody familiar to Rivière's ears. Then he pulled up the aërial and stretched his limbs, yawning and smiling; another journey done.

The pilot who had just made land greeted the pilot of the Europe mail, who was lolling, his hands in his pockets, against the plane.

"Your turn to carry on?"

"Yes."

"Has the Patagonia come in?"

"We don't expect it; lost. How's the weather? Fine?"

"Very fine. Is Fabien lost then?"

They spoke few words of him, for that deep fraternity of theirs dispensed with phrases.

The transit mail-bags from Asuncion were loaded into the Europe mail while the pilot, his head bent back and shoulders pressed against the cockpit, stood motionless, watching the stars. He felt a vast power stirring in him and a potent joy.

"Loaded?" some one asked. "Then, contact!"

The pilot did not move. His engine was started. Now he would feel in his shoulders that pressed upon it the airplane come to life. At last, after all those false alarms—to start or not to start—his mind was easy. His lips were parted and in the moon his keen white teeth glittered like a jungle cub's.

"Watch out! The night, you know . . . .!"

He did not hear his comrade's warning. His hands thrust in his pockets and head bent back, he stared toward the clouds, mountains and seas and rivers, and laughed silently. Soft laughter that rustled through him like a breeze across a tree, and all his body thrilled with it. Soft laughter, yet stronger, stronger far, than all those clouds and mountains, seas and rivers.

"What's the joke?"

"It's that damned fool Rivière, who said . . . who thinks I've got the wind up!"

# XXIII

In a minute he would be leaving Buenos Aires and Rivière, on active service once again, wanted to hear him go. To hear his thunder rise and swell and die into the distance like the tramp of armies marching in the stars.

With folded arms Rivière passed among the clerks and halted at a window to muse and listen. If he had held up even one departure, that would be an end of night flights. But, by launching this other mail into the darkness, Rivière had forestalled the weaklings who to-morrow would disclaim him.

Victory, defeat—the words were meaningless. Life lies behind these symbols and life is ever bringing new symbols into being. One nation is weakened by a victory, another finds new forces in defeat. To-night's defeat conveyed perhaps a lesson which would speed the coming of final victory. The work in progress was all that mattered.

Within five minutes the radio stations would broadcast the news along the line and across a thousand miles the vibrant force of life give pause to every problem.

Already a deep organ-note was booming; the plane.

Rivière went back to his work and, as he passed, the clerks quailed under his stern eyes; Rivière the Great, Rivière the Conqueror, bearing his heavy load of victory.

# FLIGHT TO ARRAS

*Translated from the French*
*By Lewis Galantière*

# I

Surely I must be dreaming. It is as if I were fifteen again. I am back at school. My mind is on my geometry problem. Leaning over the worn black desk, I work away dutifully with compass and ruler and protractor. I am quiet and industrious.

Near by sit some of my schoolmates, talking in murmurs. One of them stands at a blackboard chalking up figures. Others less studious are playing bridge. Out-of-doors I see the branch of a tree swaying in the breeze. I drop my work and stare at it. From an industrious pupil I have become an idle one. The shining sun fills me with peace. I inhale with delight the childhood odor of the wooden desk, the chalk, the blackboard in this schoolhouse in which we are quartered. I revel in the sense of security born of this daydream of a sheltered childhood.

What course life takes, we all know. We are children, we are sent to school, we make friends, we go to college—and we are graduated. Some sort of diploma is handed to us, and our hearts pound as we are ushered across a certain threshold, marched through a certain porch, the other side of which we are of a sudden grown men. Now our footfalls strike the ground with a new assurance. We have begun to make our way in life, to take the first few steps of our way in life. We are about to measure our strength against real adversaries. The ruler, the T square, the compass have become weapons with which we shall build a world, triumph over an enemy. Playtime is over.

All this I see as I stare at the swaying branch. And I see too that schoolboys have no fear of facing life. They champ at the bit. The jealousies, the trials, the sorrows of the life of man do not intimidate the schoolboy.

But what a strange schoolboy I am! I sit in this schoolroom, a schoolboy conscious of my good fortune and in no hurry to face life. A schoolboy aware of its cares. . . .

Dutertre comes by, and I stop him.

"Sit down. I'll do some card-tricks for you."

Dutertre sits facing me on a desk as worn as mine. I can see his dangling legs as he shuffles the cards. How pleased with myself I am when I pick out the card he has in mind! He laughs. Modestly, I smile. Pénicot comes up and puts his arm across my shoulder.

"What do you say, old boy?"

How tenderly peaceful all this is!

A school usher—is it an usher?—opens the door and summons two among us. They drop their ruler, drop their compass, get up, and go out. We follow them with our eyes. Their schooldays are over. They have been released for the business of life. What they have learnt, they are now to make use of. Like grown men, they are about to try out against other men the formulas they have worked out.

Strange school, this, where each goes forth alone in turn. And without a word of farewell. Those two who have just gone through the door did not so much as glance at us who remain behind. And yet the hazard of life, it may be, will transport them farther away than China. So much farther! When schooldays are past, and life has scattered you, who can swear that you will meet again?

The rest of us, those still nestling in the cosy warmth of our incubator, go back to our murmured talk.

"Look here, Dutertre. To-night—."

But once again the same door has opened. And like a court sentence the words ring out in the quiet schoolroom:

"Captain de Saint-Exupéry and Lieutenant Dutertre report to the major!"

Schooldays are over. Life has begun.

"Did you know it was our turn?"

"Pénicot flew this morning."

"Oh, yes."

The fact that we had been sent for meant that we were to be ordered out on a sortie. We had reached the last days of May, 1940, a time of full retreat, of full disaster. Crew after crew was being offered up as a sacrifice. It was as if you dashed glassfuls of water into a forest fire in the hope of putting it out. The last thing that could occur to any one in this world that was tumbling round our ears was the notion of risk or danger. Fifty reconnaissance crews was all we had for the whole French army. Fifty crews of three men each—pilot, observer, and gunner. Out of the fifty, twenty-three made up our unit—Group 2-33. In three weeks, seventeen of the twenty-three had vanished. Our Group had melted like a lump of wax. Yesterday, speaking to Lieutenant Gavoille, I had let drop the words, "Oh, we'll see about that when the war is over." And Gavoille had answered, "I hope you don't mean, Captain, that you expect to come out of the war alive?"

Gavoille was not joking. He was sincerely shocked. We knew perfectly well that there was nothing for us but to go on flinging ourselves into the forest fire. Even though it serve no purpose. Fifty crews for the whole of France. The whole strategy of the French army rested upon our shoulders. An immense forest fire raging, and a hope that it might be put out by the sacrifice of a few glassfuls of water. They would be sacrificed.

And this was as it should be. Who ever thought of complaining? When did anyone ever hear, among us, anything else than

"Very good, sir. Yes, sir. Thank you, sir. Quite right, sir." Throughout the closing days of the French campaign one impression dominated all others—an impression of absurdity. Everything was cracking up all round us. Everything was caving in. The collapse was so entire that death itself seemed to us absurd. Death, in such a tumult, had ceased to count. But we ourselves did not count.

Dutertre and I went into the major's office. The major's name was Alias. As I write, he is still in command of Group 2-33, at Tunis.

"Afternoon, Saint-Ex. Hello, Dutertre. Sit down."

We sat down. The major spread out a map on the table and turned to his clerk:

"Fetch me the weather reports."

He sat tapping on the table with his pencil. I stared at him. His face was drawn. He had had no sleep. Back and forth in a motorcar, he had driven all night in search of a phantom General Staff. He had been summoned to division headquarters. To brigade headquarters. He had argued and wrangled with supply depots that never delivered the spare parts they had promised. His car had been bottled up in the crazy traffic. He had supervised our last moving out and our most recent moving in—for we were driven by the enemy from one field to another like poor devils scrambling in the van of a relentless bailiff. Alias had succeeded in saving our planes, saving our lorries, saving the stores and files of the Group. He looked as if he had reached the end of his strength, of his nerves.

"Well," he said, and he went on tapping with his pencil. He was still not looking at us.

A moment passed before he spoke again. "It's damned awkward," he said finally; and he shrugged his shoulders. "A damned awkward sortie. But the Staff want it done. They very much want it done. I argued with them; but they want it done. . . . And that's that."

Dutertre and I sat looking out of the window. Here too a branch was swaying in the breeze. I could hear the cackle of the hens. Our Intelligence Room had been set up in a schoolhouse; the major's office was in a farmhouse.

It would be easy to write a couple of fraudulent pages out of the contrast between this shining spring day, the ripening fruit, the chicks filling plumply out in the barnyard, the rising wheat—and death at our elbow. I shall not write that couple of pages because I see no reason why the peace of a spring day should constitute a contradiction of the idea of death. Why should the sweetness of life be a matter for irony?

But a vague notion did go through my mind as I stared out of Alias' window. "The spring has broken down," I said to myself. "The season is out of order." I had flown over abandoned threshing machines, abandoned binders. I had seen motorcars deserted in roadside ditches. I had come upon a village square standing under water while the village faucet—"the fountain" as our people call it—stood open and the stream flowed on.

And suddenly a completely ridiculous image came into my mind. I thought of clocks out of order. All the clocks of France—out of order. Clocks in their church steeples. Clocks on railway stations. Chimney clocks in empty houses. A charnelhouse of clocks. "The war," I said to myself, "is that thing in which clocks are no longer wound up. In which beets are no longer gathered in. In which farm carts are no longer greased. And that water, collected and piped to quench men's thirst and to whiten the Sunday laces of the village women—that water stands now in a pool flooding the square before the village church."

As for Alias, he was talking like a bedside physician. "Hm," says the doctor with a shake of the head, "rather awkward, this"; and you know that he is hinting that you ought to be making your will, thinking of those you are about to leave behind. There was no question in Dutertre's mind or mine that Alias was talking about sacrificing another crew.

"And," Alias went on, "things being as they are, it's no good worrying about the chances you run."

Quite so. No good at all. And it's no one's fault. It's not our fault that we feel none too cheerful. Not the major's fault that he is ill at ease with us. Not the Staff's fault that it gives orders. The major is out of sorts because the orders are absurd. We know that they are absurd; but the Staff knows that as well as we do. It gives orders because orders have to be given. Giving orders is its trade, in time of war. And everyone knows what war looks like. Handsome horsemen transmit the orders—or rather, to be modern about it, motorcyclists. The orders ordain events, change the face of the world. The handsome horsemen are like the stars—they bring tidings of the future. In the midst of turmoil and despair, orders arrive, flung to the troops from the backs of steaming horses. And then all is well—at least, so says the blueprint of war. So says the pretty picture-book of war. Everybody struggles as hard as he can to make war look like war. Piously respects the rules of the game. So that war may perhaps be good enough to agree to look like war.

Orders are given for the sacrifice of the air arm because war must be made to look like war. And nobody admits meanwhile that this war looks like nothing at all. That no part of it makes sense. That not a single blueprint fits the circumstances. That the puppets have been cut free of the strings which continue to be pulled.

In all seriousness the Staffs issue orders that never reach anybody. They ask us for intelligence impossible to provide. But the air arm cannot undertake to explain war to the Staffs. Reconnaissance pilots might be able to test or verify the Staffs' hypotheses. But there are no longer any hypotheses. Fifty reconnaissance crews are asked to sketch the face of a war that has no face. The Staffs appeal to us as if we were a tribe of fortune-tellers.

While Alias was speaking I threw a glance at Dutertre—my fortune-telling observer. This was what he said afterwards.

"What do they take us for, sending us off on low-altitude sorties? Only yesterday I had to tick off a colonel from division headquarters who was talking the same rot. 'Will you tell me,' I said to him; 'will you tell me how I am going to report the enemy's position to you from an altitude of fifty feet when I'm doing three hundred miles an hour?' He looked at me as if I was the one who was mad. 'Why,' he said, 'that's easy. You can tell according to whether they shoot or not. If they shoot at you, the positions are German.' Imagine! The bloody fool!"

What Dutertre knew, and the colonel seemed to forget, was that the French army never saw French aeroplanes. We had roughly one thousand planes scattered between Dunkerque and Alsace. Diluted in infinity, so far as the men on the ground were concerned. The result was that when a plane roared across our lines, it was virtually certain to be a German. You let fly with all the anti-aircraft guns you had even before you saw him, the instant you heard him; for otherwise he had dropped his bombs and was off before you could say "wink!"

"A precious lot of intelligence we'll bring home working their way!" Dutertre said.

Of course they take our intelligence into account, since the blueprint of war requires that intelligence officers make use of intelligence. But even their war-by-the-blueprint had broken down. We knew perfectly well that they would never be able to make use of our intelligence—luckily. It might be brought back by us; but it would never be transmitted to the Staff. The roads would be jammed. The telephone lines would be cut. The Staff would have moved in a hurry. The really important intelligence—the enemy's position—would have been furnished by the enemy himself.

For example. A few days earlier we of Group 2-33, having been driven back by successive stages to the vicinity of Laon, were wondering how near the front might now be—how soon we should be forced to move again. A lieutenant was sent off for information to the general in command who was seven miles away. Halfway between the airfield and the general's headquarters the lieutenant's motorcar ran up against a steam-roller behind which two armored cars were hidden. The lieutenant made a U-turn and started away, but a blast of machine-gun fire killed him instantly and wounded his chauffeur. The armored cars were German. They taught us where the "front" was.

The General Staff was like a first-rate bridge player who is asked by someone sitting in a game in the next room, "What do you think I ought to do with the queen of spades?" How can the expert, knowing nothing of that particular game, have an opinion about that queen of spades?

Actually, a General Staff has no right to be without an opinion. Besides, so long as certain elements are still in its hands, it is bound to make use of them—since otherwise it will lose its control over them. The opponents will work a squeeze play. Thus, the General Staff must take risks. So long as there is a war on it must act, even though it act blindly.

But it is, nevertheless, very hard to say what shall be done with the queen of spades when you haven't a hand in the game. What we had learnt, meanwhile—at first with surprise, and then with the feeling that we ought to have seen it coming—was that once the cracking up begins, the machine stops running. There is no soldiering for the soldier to do.

You might think that in retreat and disaster there ought to be such a flood of pressing problems that one could hardly decide which to tackle first. The truth is that for a defeated army the problems themselves vanish. I mean by this that a defeated army no longer has a hand in the game. What is one to do with a plane, a tank, in short a queen of spades, that is not part of any known game? You hold the card back; you hesitate; you rack your brains to find use for it—and then you fling it down on the chance that it may take a trick.

Commonly, people believe that defeat is characterized by a general bustle and a feverish rush. Bustle and rush are the signs of victory, not of defeat. Victory is a thing of action. It is a house in the act of being built. Every participant in victory sweats and puffs, carrying the stones for the building of the house. But defeat is a thing of weariness, of incoherence, of boredom. And above all of futility.

For in the first place these sorties on which we were sent off were futile. More murderous and more futile with every day that passed. Against the avalanche that was overwhelming them our generals could defend themselves only with what they had. They had to fling down their trumps; and Dutertre and I, as we sat listening to the major, were their trumps.

The major was sketching for us the afternoon's program. He was sending us off to fly a photography sortie at thirty thousand feet and thereafter to do a reconnaissance job at two thousand feet above the German tank parks scattered over a considerable area round Arras. His voice was as deliberate as if he were saying, "and then you take the second street on the right to a square where you will see a tobacco shop."

What could we answer but "Very good, sir?" The sortie was as futile as that—the language as lyrical as the futility of the sortie required.

I had my own thoughts. "Another crew flung away," I said to myself. My head was buzzing, buzzing with many things; but I said to myself that I'd wait. If we got back, if we were alive that night, I'd do my thinking then.

If we were alive. When a sortie was not "awkward," one plane out of three got back. Naturally, the ratio was not the same when the sortie was a nasty one. But I was not weighing my chances of getting back. Sitting there in the major's office, death seemed to me neither august, nor majestic, nor heroic, nor poignant. Death seemed to me a mere sign of disorder. A consequence of disorder. The Group was to lose us more or less as baggage becomes lost in the hubbub of changing trains.

Not that on the subject of war, of death, of sacrifice, of France I do not think quite other things than what I now say; but sitting in that office my thoughts were without a compass, my language was a blur. I sat thinking in contradictions. My concept of truth had been shattered, and the best I could do was to stare at one fragment after another. "If I am alive," I said to myself, "I shall do my thinking tonight." Night, the beloved. Night, when words fade and things come alive. When the destructive analysis of day is done, and all that is truly important becomes whole and sound again. When man reassembles his fragmentary self and grows with the calm of a tree.

Day belongs to family quarrels, but with the night he who has quarrelled finds love again. For love is greater than any wind of words. And man, leaning at his window under the stars, is once again responsible for the bread of the day to come, for the slumber of the wife who lies by his side, all fragile and delicate and contingent. Love is not thinking, but being. As I sat facing Alias I longed for night and for the rebirth in me of the being that merits love. For night, when my thoughts would be of civilization, of the destiny of man, of the savor of friendship in my native land. For night, so that I might yearn to serve some overwhelming purpose which at this moment I cannot define. For night, so that I might perhaps advance a step towards fixing it in my unmanageable language. I longed for night as the poet might do, the true poet who feels himself inhabited by a thing obscure but powerful, and who strives to erect images like ramparts round that thing in order to capture it. To capture it in a snare of images.

And as I sat there longing for night, I was for the moment like a Christian abandoned by grace. I was about to do my job with Dutertre honorably, that was certain. But to do it as one honors ancient rites when they have no longer any significance. When the god that lived in them has withdrawn from them. I should wait for night, I said to myself; and if I was still alive I would walk alone on the highway that runs through our village. Alone and safely isolated in my beloved solitude. So that I might discover why it is I ought to die.

## II

I awoke out of my daydream—was startled out of it by an astonishing proposal.

"If this sortie bothers you, Saint-Ex; if you don't feel up to it today, I can—."

"Oh, come, Major!"

He knew perfectly well that his proposal was idiotic. And I knew why he made it. If a pilot doesn't get back you begin to recall how solemn he was when he was ordered out. You say to yourself that he must have had a premonition of his end. And you accuse yourself of having wilfully brushed it aside. You take time out for an attack of conscience.

The major's scruple reminded me of Israel. Two days before, I had been sitting smoking at the window of the Intelligence Room. Israel, when I caught sight of him through the window, was walking swiftly past. His nose was red. A big nose, very Jewish and very red. Suddenly there seemed to me something queer about that big red nose.

This Israel, whose nose I was staring at, was a man I profoundly liked. He was one of the most courageous pilots of the Group. One of the most courageous and one of the most modest. He had heard so much talk of Jewish craftiness that he probably mistook his courage for a form of craftiness. To gain a victory is to act craftily.

There I sat, watching that red nose that gleamed in my sight only for an instant, so swift were the steps that carried Israel and his nose out of view. I turned to Gavoille, and without meaning to make a joke of it, I said:

"Why do you suppose his nose is like that?"

Gavoille answered: "His mother made it like that." And then added quickly: "Low-altitude sortie. Can't blame the fellow."

That night, when we had given up looking for Israel to get back, I thought again of that nose, planted in the middle of a totally expressionless face and yet revealing, with a sort of genius of its own, the burden of the thoughts revolving in the man's mind. If it had been my job to order Israel on that sortie, the memory of his nose would have haunted me like a reproach. Israel, surely, had responded to the order with no more than a "Yes, sir," a "Very good, sir." Israel, surely, had not allowed a single muscle of his face to quiver on hearing the order. But gently, insidiously, treacherously, his nose had reddened. Israel had been able to control the muscles of his face, but not the color of his nose. And in the silence in which he had received the order, his nose had taken advantage of him. Unknown to Israel, it had made clear to the major its emphatic disapproval of the sortie.

This was the kind of thing that made Alias hesitate to send into action men he imagined might be subject to premonitions. Premonitions are more often false than true; but when you are seized by one, a military order will sound like a court sentence. And Alias was not a judge, after all, but a group commander.

There was the case the other day of the gunner I shall call T. As Israel was all courage, so T. was all fear. He is the only man I have ever known who really felt fear. When, during the war, you gave T. an order you released in him at that moment a wave of dizziness. Something simple, relentless, and gradual. Rising slowly from his feet to his head, a stiffening would come over his whole body. Little by little his face would go totally blank. And his eyes would begin to shine.

Unlike Israel, whose nose, reddened with irritation, had seemed to me so dejected at the thought of the probable death of Israel, no psychic mutation took place in T. He did not react, he moulted. When you had finished giving T. an order you discovered that you had lit a flame of anguish in him, and that the anguish had begun to spread a sort of even glow through his being. Thereafter, nothing at all could reach him. You felt in the man the gradual spread of a desert of indifference that intervened between him and the universe. Never in any other man on earth have I perceived this form of ecstasy.

"I shouldn't have let him fly that day," Alias said to me later.

For that day, when the major had given T. his orders, T. had not merely turned white, he had begun to smile. Quite plainly to smile. Probably as tortured men smile when, really, the executioner has gone too far.

"You're off your feed today, T. I'll get another gunner."

"If you please, sir. It's my turn," T. had answered. He was standing respectfully at attention, eyes front and perfectly motionless.

"Still, if you don't feel sure of yourself—."

"It's my turn out, sir."

"Come, T., look here—."

"Sir!" T. had interrupted; and his whole body looked carved out of rock.

"So," Alias concluded, "I let him have his way."

Exactly what happened, we never knew. T., sitting aft as gunner of the crew, had seen a German fighter bear down on him. The German's guns had jammed, and he had turned tail and vanished. T. had exchanged remarks with his pilot through the speaking tube all the way back to the neighborhood of their base. The pilot had observed nothing abnormal in T.'s conversation. But about five minutes before landing T. had stopped talking, and the pilot had been unable to raise him.

That same evening, T. was brought in, his skull split open by the tail-unit of his own plane. He had tried to bail out over home territory where he was completely out of danger. The plane had been flying at high speed, and he had done a bad job of parachuting. The passage of that German fighter had been irresistible, a siren call.

"Better get along and dress, now," the major said. "I want you off the ground at five-thirty."

We said, "See you this evening, sir," and the major responded by a vague wave of the hand. Was it superstition? I turned to leave, became aware that my cigarette was out, and was fumbling in vain through all my pockets when the major said testily:

"Why is it you never carry any matches?"

It was true; and with this substitute for "Good luck!" in my ears I shut the door saying to myself, "Why is it I never have a match on me?"

Dutertre said, "This sortie has got on his nerves."

He doesn't give a damn about it, I thought. But I didn't say so aloud, for I wasn't thinking of Alias. I was thinking of man in general. I had been brought up with a jerk by a very evident fact which men do not trouble to see—that the life of the spirit, the veritable life, is intermittent, and only the life of the mind is constant. This instant and spontaneous reflection leads back to Alias in a roundabout way.

Man's spirit is not concerned with objects; that is the business of our analytical faculties. Man's spirit is concerned with the significance that relates objects to one another. With their totality, which only the piercing eye of the spirit can perceive. The spirit, meanwhile, alternates between total vision and absolute blindness. Here is a man, for example, who loves his farm—but there are moments when he sees in it only a collection of unrelated objects. Here is a man who loves his wife—but there are moments when he sees in love nothing but burdens, hindrances, constraints. Here is a man who loves music—but there are moments when it cannot reach him. What we call a nation is certainly not the sum of the regions, customs, cities, farms, and the rest that man's intelligence is able at any moment to add up. It is a Being. But there are moments when I find myself blind to beings—even to the being called France.

Major Alias had spent the previous night at Staff headquarters discussing what was in effect pure logic. Pure logic is the ruin of the spirit. Afterwards he had driven back, and driving back he had worn himself out getting through the tangled traffic. Having finally reached his billet he had found a hundred details to look after, those details that fray a man's nerves and set him on edge. And this afternoon he had sent for us and ordered us to embark upon an utterly impossible sortie. What were we to him? Particles in the universal chaos. We were not Saint-Exupéry and Dutertre to him—each with our own way of seeing or not seeing things, of thinking, walking, smiling, drinking. We were mere details in a vast structure to see the whole of which demanded more time, more silence, more perspective than he could possibly obtain. Had my face been afflicted with a tic, he would have been able to see nothing but the tic. He would have sent out over Arras the memory of a tic. In this senseless hullabaloo, in this avalanche, we ourselves, each of us, saw nothing but particles. That voice. That nose. That tic. And particles are not the objects of anybody's emotion.

Thus I was not thinking about Alias specifically, but about man in general. A friend you love has died, and it is you who must see that he is decently buried. At that moment you have no contact with your dead friend. How can you have? Death is a thing of grandeur. It brings instantly into being a whole new network of relations between you and the ideas, the desires, the habits of the man now dead. It is a rearrangement of the world. Nothing has changed visibly, yet everything has changed. The pages of the book are the same, but the meaning of the book is different. And how can you, who are busy with funeral details, know any of this? Do you wish to bring the dead friend to mind? You must be able to imagine yourself needing him. At that moment you will miss him. Imagine him needing you. Ah, but he no longer needs you!

Imagine those Wednesdays when, invariably, you lunched together. Wednesday is now a vacuum. Life, we know, has to be seen in perspective. But on a day of burial there is no perspective—for space itself is annihilated. Your dead friend is still a fragmentary being. The day you bury him is a day of chores and crowds, of hands false or true to be shaken, of the immediate cares of mourning. The dead friend will not really die until tomorrow, when silence is round you again. Then he will show himself complete, as he was—to tear himself away, as he was, from the substantial you. Only then will you cry out because of him who is leaving and whom you cannot detain.

I am still on the track of my thought when I say that I do not like the pretty picture-book of war. The gruff warrior squeezing back a tear and hiding his honest emotion under a grumpy exterior. What nonsense! The gruff warrior is not hiding anything at all. If he lets fly a gruff remark it is because a gruff remark has come into his mind.

Nor does it matter for my purpose whether a man be decent or a brute. Major Alias is a sensitive person. If Dutertre and I fail to get back it will probably affect him more than anyone else in the Group. Provided, however, that he think of Saint-Exupéry and Dutertre, and not of a sum of unrelated particles. Provided that he be allowed the silence in which to effect this reconstruction of ourselves. For if, tonight, the baliff at our heels once more constrains the Group to move, a single broken-down lorry will suffice to put off our death until another time. Alias will forget to be affected by our death.

The life of the spirit, I say, is intermittent. My own spirit as much as Alias'. I am off on an "awkward" sortie. Is my mind filled with the thought of the war of the Nazi against the Occident? Not at all. I think in terms of immediate details. I think of possible wounds. I think of the absurdity of flying over German-held Arras at two thousand feet. Of the futility of the intelligence we are asked to bring back. Of the interminable time it takes to dress in these clothes that remind me of men made ready for the executioner. And I think of my gloves. Where the devil are my gloves? I have lost my gloves.

I can no longer see the cathedral in which I live. I am dressing for the service of a dead god.

## III

"Get going! Where are my gloves? . . . No, not those. Have a look in my bag."

"Sorry, sir. Can't find them."

"God, you're a fool!"

Everybody is a fool. My orderly, who doesn't know where my gloves are. Hitler, who unloosed this mad war. And that fellow on the General Staff, obsessed by low-altitude sorties.

"I asked you to get me a pencil. I have been asking you for ten minutes to find me a pencil. Haven't you got a pencil?"

"Here it is, sir."

One man, at least, who is not a fool.

"Tie a string round it. Now knot the string through this buttonhole. . . . I say, gunner, you seem to be taking things very easily."

"I'm all ready, sir."

"Oh!"

And my observer. I swung round to him.

"Everything shipshape, Dutertre? Nothing missing? Worked out your course?"

He has worked out his course. "Awkward" sortie indeed! Where is the sense, I ask you, in sending a crew out to be murdered for the sake of intelligence that is sure to be useless and will never reach the Staff anyway, even if one of us lives to report it?

"Mediums," I said aloud. "They must have a crew of mediums on the General Staff."

"What do you mean, Captain?"

"How do *you* think we'll report to them? They are going to communicate with us. Table tipping. Automatic writing."

Not very funny; but I went on grousing.

"General Staffs! Let them fly their own damned sorties!"

It takes a long time to dress for a sortie that you know is a hopeless one. A long time to harness yourself only for the fun of being blasted to bits. There are three thicknesses of clothing to be put on, one over the other: that takes time. And this clutter of accessories that you carry about like an itinerant pedlar! All this complication of oxygen tubes, heating equipment; these speaking tubes that form the "inter-com" running between the members of the crew. This mask through which I breathe. I am attached to the plane by a rubber tube as indispensable as an umbilical cord. The plane is plugged in to the circulation of my blood. Organs have been added to my being, and they seem to intervene between me and my heart. From one minute to the next I grow heavier, more cumbrous, harder to handle. I turn round all of a piece, and when I bend down to tighten my straps or pull at buckles that resist, all my joints creak aloud. My old fractures begin to hurt again.

"Hand me another helmet. I've told you twenty times that my own won't do. It's too tight."

God knows whv. but a man's skull swells at high altitude. A

helmet that fits perfectly on the ground becomes a vise pressing on the skull at thirty thousand feet.

"But this is another helmet, sir. I sent back your old one."

"Huh!"

I cannot stop grousing, and I grouse without remorse. A lot of good it does! Not that it is important. This is the moment of timelessness. This is the crossing of the inner desert of anguish. There is no god here. There is no face to love. There is no France, no Europe, no civilization. There are particles, detritus, nothing more. I feel no shame at this moment in praying for a miracle that should change the course of this afternoon. The miracle, for instance, of a speaking tube out of order. Speaking tubes are always going out of order. Trashy stuff! A speaking tube out of order would preserve us from the holocaust.

Captain Vezin came in with a gloomy look. No pilot ever got off the ground without a dose of Captain Vezin's gloom. His job was to report upon the position of the Germain air outposts. To tell us where they were. Vezin is my friend and I am very fond of him; but he is a bird of ill omen. I prefer not to meet him when I am about to take off.

"Looks bad, old boy," said Vezin. "Very bad. Very bad indeed."

And didn't he pull a sheaf of papers out of his pocket, to impress me! Then, looking as me suspiciously, he said:

"How are you going out?"

"By the town of Albert."

"I thought so. I knew it. Bad business."

"Stop talking like a bloody fool! What's up?"

"You'll never make it. You'll have to give up this sortie."

Give up this sortie! Very kind of him to say so. Let him tell that to God the Father. Perhaps He'll put a curse on our speaking tubes.

"You'll never get through, I tell you."

"And why will I never get through?"

"Because there are three groups of German fighters circling permanently over Albert. One at eighteen thousand feet, another at twenty-five thousand, and a third at thirty-three thousand. They fly in relays and hang on until they are relieved. It's what I call *categorically blocked*. You'll fly into a German net. See here. . . ."

He shoved a sheet of paper at me on which he had scribbled an absolutely unintelligible demonstration of his argument.

Vezin had done much better to keep his nose out of my affairs. His pompous *categorically blocked* had impressed me, confound him! I thought instantly of red lights and traffic tickets. Only, this was a place where a ticket meant death. It was his *categorically* that particularly galled me. It seemed to be aimed at me personally.

I made a great effort to think clearly. "The enemy," I said to myself, "always defends his position *categorically*. Damned nonsense, these big words! And besides, why should I worry about German fighter planes? At thirty thousand feet they would get me before I so much as suspected their presence, and at two thousand feet it was the anti-aircraft that would bring me down, not the fighters. It couldn't possibly miss me." Suddenly I became belligerent.

"In short, what you're telling me is that the Germans have an air force, and therefore my sortie is not altogether advisable. Run along and tell that to the General."

It wouldn't have cost Vezin anything to reassure me pleasantly, instead of upsetting me. Why couldn't he have said, "Oh, by the way. The Germans have a few fighters aloft over Albert"?

It would have come to the same thing.

We climbed in. I had still to test the inter-com.

"Can you hear me, Dutertre?"

"I hear you, Captain."

"You, gunner! Hear me?"

"I . . . Yes, sir. Clearly."

"Dutertre! Can you hear the gunner?"

"Clearly, Captain."

"Gunner! Can you hear Lieutenant Dutertre?"

"I . . . er . . . Yes, sir. Clearly."

"What makes you stutter back there? What are you hesitating about?"

"Sorry, sir. I was looking for my pencil."

The speaking tubes were not out of order.

"Gunner! Have a look at your oxygen bottles. Air-pressure normal?"

"I . . . Yes, sir. Normal."

"In all three bottles?"

"All three, sir."

"All set, Dutertre?"

"All set, Captain."

"All set, gunner?"

"All set, sir."

We took off.

## IV

Human anguish is the product of the loss by man of his true identity. I sit waiting for a telegram which is to announce to me either a death or a recovery. Time flows by unutilized and holds me in suspense. Time has ceased to be a stream that feeds me, nourishes me, adds growth to me as to a tree. The man that I shall be when the news comes, dwells outside me: he is moving towards me like a ghost about to fuse with me. And for want of knowing who I am, I am suspended in anguish. The bad news, when it comes, puts an end to my suspense. It causes me to suffer, which is not the same thing.

T. never knew whether, in the hour to come, he was to be transmuted into a living man or a dead man. He was aware of only one thing—the flow of time, running like sand through his fingers while he waited for the coming of a certain instant too rich in power for his resistance.

For me, piloting my plane, time has ceased to run sterile through my fingers. Now, finally, I am installed in my function. Time is no longer a thing apart from me. I have stopped projecting myself into the future. I am no longer he who may perhaps dive down the sky in a vortex of flame. The future is no longer a haunting phantom, for from this moment on I shall myself create the future by my own successive acts. I am he who checks the course and holds the compass at 313°. Who controls the revolutions of the propeller and the temperature of the oil. These are healthy and immediate cares. These are household cares, the little duties of the day that take away the sense of growing older. The day becomes a house brilliantly clean, a floor well waxed, oxygen prudently doled out. . . . Thinking which, I check the oxygen flow, for we have been rising fast and are at twenty-two thousand feet already.

"Oxygen all right, Dutertre? How do you feel?"

"First-rate, Captain."

"You, gunner! How's your oxygen?"

"I . . . er . . . Shipshape, sir."

"Haven't you found that pencil yet?"

And I am he who checks his machine guns, putting a finger on button S, on button A. . . . Which reminds me.

"Gunner! No good-sized town behind you, in your cone of fire?"

"Er . . . all clear, sir."

"Check your guns. Let fly."

I hear the blast of the guns.

"Work all right?"

"Worked fine, sir."

"All of them?"

"Er . . . yes, sir. All of them."

I test my own and wonder what becomes of all the bullets that we scatter so heedlessly over our home territory. They never kill any one. The earth is vast.

Now time is nourishing me with every minute that passes. I am a thing as little the prey of anguish as a ripening fruit. Of course the circumstances of this flight will change round me. The circumstances and the problems. But I dwell now well inside the fabrication of the future. Time, little by little, is kneading me into shape. A child is not frightened at the thought of being patiently transmuted into an old man. He is a child and he plays like a child. I too play my games. I count the dials, the levers, the buttons, the knobs of my kingdom. I count one hundred and three objects to check, pull, turn, or press. (Perhaps I have cheated in counting my machine-gun controls at two—one for the fire-button, and another for the safety-catch.) Tonight when I get back I shall amaze the farmer with whom I am billeted. I shall say to him:

"Do you know how many instruments a pilot has to keep his eye on?"

"How do you expect me to know that?"

"No matter. Guess. Name a figure."

"What figure?"

My farmer is not a man of tact.

"Any figure. Name one."

"Seven."

"One hundred and three!"

And I shall smile with satisfaction.

Another thing contributes to my peace of mind—it is that all the instruments that were an encumbrance while I was dressing have now settled into place and acquired meaning. All that tangle of tubes and wiring has become a circulatory network. I am an organism integrated into the plane. I turn this switch, which gradually heats up my overall and my oxygen, and the plane begins to generate my comfort. The oxygen, incidentally, is too hot. It burns my nose. A complicated mechanism releases it in proportion to the altitude at which I fly, and I am flying high. The plane is my wet-nurse. Before we took off, this thought seemed to me inhuman; but now, suckled by the plane itself, I feel a sort of filial affection for it. The affection of a nursling.

My weight, meanwhile, is comfortably distributed over a variety of points of support. I am like a feeble convalescent stripped of bodily consciousness and lying in a chaise-longue. The convalescent exists only as a frail thought. My triple thickness of clothing is without weight in my seat. My parachute, slung behind, lies against the back of my seat. My enormous boots rest on the bar that operates the rudder. My hands that are so awkward when first I slip on the thick stiff gloves, handle the wheel with ease. Handle the wheel. Handle the wheel. . . .

"Dutertre!"

". . . t'n?"

"Something's wrong with the inter-com. I can't hear you. Check your contacts."

"I can . . . you . . . ctly."

"Shake it up! Can you still hear me?"

Dutertre's voice came through clearly.

"Hear you perfectly, Captain."

"Good! Dutertre, the confounded controls are frozen again. The wheel is stiff and the rudder is stuck fast."

"That's great! What altitude?"

"Thirty-two thousand."

"Temperature?"

"Fifty-five below zero. How's your oxygen?"

"Coming fine."

"Gunner! How's your oxygen?"

No answer.

"Hi! Gunner!"

No answer.

"Do you hear the gunner, Dutertre?"

"No."

"Call him."

"Gunner! Gunner!"

No answer.

"He must have passed out, Captain. We shall have to dive."

I didn't want to dive unless I had to. The gunner might have dropped off to sleep. I shook up the plane as roughly as I could.

"Captain, sir?"

"That you, gunner?"

"I . . . er . . . yes, sir."

"Not sure it's you?"

"Yes, sir."

"Why the devil didn't you answer before?"

"I had pulled the plug, sir. I was testing the radio."

"You're a bloody fool! Do you think you're alone in this plane? I was just about to dive. I thought you were dead."

"Er . . . no, sir."

"I'll take your word for it. But don't play that trick on me again. Damn it! Let me know before you cut."

"Sorry, sir. I will. I'll let you know, sir."

Had his oxygen flow stopped working, he wouldn't have known it. The human body receives no warning. A vague swooning comes over you. In a few seconds you have fainted. In a few minutes you are dead. The flow has constantly to be tested—particularly by the pilot. I pinched my tube lightly a few times and felt the warm life-bringing puffs blow round my nose.

It came to this, that I was working at my trade. All that I felt was the physical pleasure of going through gestures that meant something and were sufficient unto themselves. I was conscious neither of great danger (it had been different while I was dressing) nor of performing a great duty. At this moment the battle between the Nazi and the Occident was reduced to the scale of my job, of my manipulation of certain switches, levers, taps. This was as it should be. The sexton's love of his God becomes a love of lighting candles. The sexton moves with deliberate step through a church of which he is barely conscious, happy to see the candlesticks bloom one after the other as the result of his ministrations. When he has lighted them all, he rubs his hands. He is proud of himself.

I for my part am doing a good job of regulating the revolutions of the propeller, and the needle of my compass lies within a single degree of my course. If Dutertre happens to have his eye on the compass, he must be marvelling at me.

"I say, Dutertre! Compass on the course? How does it look?"

"Won't do, Captain. Too much drift. A little kick to starboard."

Well, well.

"Crossing our lines, Captain. I've started my camera. What's your altitude?"

"Thirty-three thousand."

## V

"Your course, Captain!"

He's right. I was drifting to port. And not by chance, either. It was the town of Albert that was putting me off. I could make it faintly out, far ahead. But already it was shouldering me off with all the weight of its *categorically blocked*. Extraordinary, the memory secreted in the recesses of the human body. My body was remembering every sudden crash of the past, every cranial fracture, each of those nights in hospital with their comas as sticky as molasses. My body is afraid of blows. It struggles to avoid Albert. The moment I leave it to itself, we drift to port. It shies left like an old horse fearful for life of the obstacle that had once frightened it. And it is really my body, not my mind that I mean. The moment my mind wanders, my body takes sly advantage of me to slip around Albert.

For it is not I who feel any anxiety. I have stopped wishing to get out of this sortie. On the ground, it had seemed to me that that was what I wanted. I had said to myself hopefully that the intercom would be out of order. I was weary, and it would be wonderful to sleep. The bed of idleness had seemed to me a magic couch. But deep down I had known perfectly well that nothing could come of getting out of this sortie except a sharp sense of discomfort. As if a necessary moulting had miscarried.

Again I was reminded of school. Of a time when I was very young. How long ago was that? I—.

"Captain!"

"What's up?"

"Er . . . nothing. I thought I had seen something."

I don't like Dutertre seeing things. . . .

Of school, yes. When you are a little boy, in boarding school, they get you up too early. They get you up at six o'clock. It is cold. You rub your eyes, and you hate class long before the bell rings. You think how wonderful it would be if you were ill and were waking up in the infirmary, where the matron would be ready with a hot cup of camomile with lots of sugar in it. The infirmary becomes a kind of paradise in your mind.

I was like that; and naturally, the first time that I caught cold I coughed much more than was called for. And I awoke in the infirmary to the sound of the bell ringing for the others. But that bell punished me for cheating. It changed me into a wraith. It rang out the passing of living hours—hours of class with its austerity, of play-time with its tumult, of the refectory with its warmth. For those who were alive, who were not, like me, in the infirmary, it sounded the realities of an enviable existence filled with jubilations, disappointments, severities, triumphs. And I lay robbed, forgotten, sick of insipid camomiles, of the sweaty bed, the blank hours.

Nothing comes of a sortie you have got out of.

Of course there are days like this when a sortie brings no satisfaction. It is too evident that we are playing a game that we call war. We are playing Cops and Robbers. We are abiding scrupulously by the rules of conduct prescribed by the history books and the rules of tactics prescribed by the war manuals. Last night, for example, I drove up to the aerodrome in a motorcar. The sentry, obedient to the rules, presented his bayonet. My car might as easily have been a German tank. We are playing at presenting bayonets to German tanks. But the tanks are playing their own game.

How can we possibly be enthusiastic about these grim charades, in which we play the part of supernumeraries, when we are asked to play on till we are killed? Death is a bit too serious for a charade. Who can dress with enthusiasm for such a part? Nobody . . . Even Hochedé who is a sort of saint, a man who has reached that state of permanent grace which surely is the final consummation of man—even Hochedé took refuge in silence. All of us dressed in silence, grumpily—and not because we were heroically modest. That grumpiness concealed no inner exaltation. It told its own story. And I knew what it meant. It was the grumpiness of an agent who is mystified by the instructions of an absentee owner, yet remains faithful to him. All of us longed for our quiet rooms, but there was not one who would really have chosen to go to bed.

For enthusiasm is not the important thing. There is no hope of enthusiasm in defeat. The important thing is to dress, climb aboard, and take off. What we ourselves think of the procedure is of no importance. A little boy in school enthusiastic about his grammar lesson would seem to me a little prig not to be trusted. The important thing is to strive towards a goal which is not immediately visible. That goal is not the concern of the mind, but of the spirit. The spirit knows how to love, but it is asleep. Talk to me about temptation! I know as much about temptation as any church father. To be tempted is to be tempted, when the spirit is asleep, to give in to the reasons of the mind.

What do I accomplish by risking my life in this mountain avalanche? I have no notion. Time and again people would say to me, "I can arrange to have you transferred here or there. That is where you belong. You will be more useful there than in a squadron. Pilots! We can train pilots by the thousand! Whereas you—." No question but that they were right. My mind agreed with them, but my instinct always prevailed over my mind.

Why was it that their reasoning never convinced me, even though I had no argument with which to defeat it? I would say to myself, "Intellectuals are kept in reserve on the shelves of the Propaganda Ministry, like pots of jam to be eaten when the war is over." Hardly an argument, I agree!

And now once again, like every other soldier of the Group, I have taken off in the face of every good reason, every obvious argument, every intellectual reflex. The moment will come when I shall know that it was reasonable to fight against reason. I have promised myself that if I am alive I shall walk alone on the highway that runs through our village. Then perhaps I shall dwell at last in my own self. And I shall see.

It may be that I shall have nothing to say about what I then see. When a woman seems to me beautiful, I have no words to say so. I see her smile, and that is all. Intellectuals take her face apart and explain it bit by bit. They do not see that smile.

To know is not to prove, nor to explain. It is to accede to vision. But if we are to have vision, we must learn to participate in the object of the vision. The apprenticeship is hard.

All day long my village was invisible to me. Before the sortie I saw in it nothing but mud walls and peasants more or less grimy. Now it is a handful of gravel thirty-three thousand feet below me. That is my village. But tonight, it may be, a watch dog will waken and bark. I have always loved the enchantment of a village dreaming aloud in the fair night by the voice of a single watch dog. And now what I ask is to see again my village tidied for sleep, its doors prudently shut upon its barns, its cattle, its customs. To see its peasants, home from the fields, their evening meal eaten and their table cleared, their children put to bed and their lamp blown out, dissolved into the silent night. And nothing more—unless perhaps, under the stiff white sheets of the countryside, the slow pulsation of their breathing, like the subsidence of a swell after a storm at sea.

God suspends the use of things and speech for the period of the nocturnal balance sheet. By the play of that irresistible slumber which loosens the fingers until morning, men will appear in my vision with open hands. And then perhaps I shall win a glimpse of that which has no name. I shall walk like the blind whose palms lead them towards the flame in the hearth. The blind cannot describe the flame, yet they have found it. Thus perhaps shall I see what it is in that dark village that we must die to protect—that which is unseen, yet like an ember beneath the ashes, lives on.

Nothing comes of a sortie you have got out of. If you are to understand a thing as simple as a village, you must first—.

"Captain!"

"Yes?"

"Six German fighters on the port bow."

The words rang in my ears like a thunderclap.

You must first. . . . You must first. . . . Ah! I do want very much to be paid off in time. I do want to have the right to love. I do want to win a glimpse of the being for whom I die.

# VI

"Gunner!"

"Sir?"

"D'you hear the lieutenant? Six German fighters. Six, on the port bow."

"I heard the lieutenant, sir."

"Dutertre! Have they seen us?"

"They have, Captain. Banking towards us. Fifteen hundred feet below us."

"Hear that, gunner? Fifteen hundred feet below us. Dutertre! How near are they?"

"Say ten seconds."

"Hear that, gunner? On our tail in a few seconds."

There they are. I see them. Tiny. A swarm of poisonous wasps.

"Gunner! They're crossing broadside. You'll see them in a second. There!"

"Don't see them yet, sir. . . . Yes, I do!"

I no longer see them myself.

"They after us?"

"After us, sir."

"Rising fast?"

"Can't say, sir. Don't think so. . . . No, sir."

Dutertre spoke. "What do you say, Captain?"

"What do you expect me to say?"

Nobody said anything. There was nothing to say. We were in God's hands. If I banked, I should narrow the space between us. Luckily, we were flying straight into the sun. At high altitude you cannot go up fifteen hundred feet higher without giving a couple of miles to your game. It was possible therefore that they migh-

lose us entirely in the sun by the time they had reached our altitude and recovered their speed.

"Still after us, gunner?"

"Still after us, sir."

"We gaining on them?"

"Well, sir. No. . . . Perhaps."

It was God's business—and the sun's.

Fighters do not fight, they murder. Still, it might turn into a fight, and I made ready for it. I pressed with both feet as hard as I could, trying to free the frozen rudder. A wave of something strange went over me. But my eyes were still on the Germans, and I bore with all my weight down upon the rigid bar.

Once again I discovered that I was in fact much less upset in this moment of action—if "action" was the word for this vain expectancy—than I had been while dressing. A kind of anger was going through me. A beneficent anger. God knows, no ecstasy of sacrifice. Rather an urge to bite hard into something.

"Gunner! Are we losing them?"

"We are losing them, sir."

Good job.

"Dutertre! Dutertre!"

"Captain?"

"I . . . nothing."

"Anything the matter?"

"Nothing. I thought. . . . Nothing."

I decided not to mention it. No good worrying them. If I went into a dive they would know it soon enough. They would know that I had gone into a dive.

It was not natural that I should be running with sweat in a temperature sixty degrees below zero. Not natural. I knew perfectly well what was happening. Gently, very gently, I was fainting.

I could see the instrument panel. Now I couldn't. My hands were losing their grip on the wheel. I hadn't even the strength to speak. I was letting myself go. So pleasant, letting oneself go. . . .

Then I squeezed the rubber tube. A gust of air blew into my nose and brought me life. The oxygen supply was not out of order! Then it must be. . . . Of course! How stupid I had been! It was the rudder. I had exerted myself like a man trying to pick up a grand piano. Flying thirty-three thousand feet in the air, I had struggled like a professional wrestler. The oxygen was being doled out to me. It was my business to use it up economically. I was paying for my orgy.

I began to inhale in swift repeated gasps. My heart beat faster and faster. It was like a faint tinkle. What good would it do to speak of it? If I went into a dive, they would know soon enough. Now I could see my instrument panel. . . . No, that wasn't true. I couldn't see it. Sitting there in my sweat, I was sad.

Life came back as gently as it had flowed out of me.

"Dutertre!"

"Captain?"

I should have liked to tell him what had happened.

"I . . . I thought . . . No."

I gave it up. Words consume oxygen too fast. Already I was out of breath. I was very weak. A convalescent.

"You were about to say something, Captain?"

"No. . . . Nothing."

"Quite sure, Captain? You puzzle me?"

I puzzle him. But I am alive.

"We are alive."

"Well, yes. For the time being."

For the time being. There was still Arras.

Thus for a minute or two I had the feeling that I should not pull through; and yet I had not observed in myself that poignant anxiety which, people say, turns the hair white in an instant. I began to think of Sagon, of what Sagon had said when, two months earlier, we had gone to see him only a few hours after he had been shot down behind our own lines. What had gone through his mind when the German fighters had surrounded him and nailed him to the stake.

## VII

I see him exactly as he was, lying in the hospital bed. His knee had been hooked and broken by the tail-unit of the plane in the course of a parachute jump, but Sagon had not felt the shock. His face and hands were rather badly burnt, but all in all Sagon's condition was not alarming. Slowly and in a matter of fact voice, as if reporting a bit of fatigue duty, he told us his story.

"I knew they had got me when I saw the air filled with tracer bullets round my plane. My instrument panel was shot to bits. Then I saw a puff of smoke forward. It wasn't much, you know. I thought it must be . . . you know . . . there's a connecting pipe. There wasn't much flame."

He stopped, and his lower lip came forward while he turned it over in his mind. It seemed to him important to be able to tell us whether the flames were high or were not high. He hesitated: "But still, flame is flame. The inter-com was working, and I told the crew they'd better jump."

In less than ten seconds a plane can turn into a torch.

"Then I opened my escape hatch. I shouldn't have done that. It let in the air . . . and the flame, you know. . . . I was sorry I'd done it."

You have a locomotive boiler spitting a torrent of flame at you, twenty thousand feet in the air, and you are sorry you've done something. I shall not play Sagon false by talking of his heroism or his modesty. He would not recognize himself in these terms. He would insist that he was sorry he had done it. As we stood round his bed it was plain that he was making a concentrated effort to be precise.

The field of consciousness is tiny. It accepts only one problem at a time. Get into a fist fight, put your mind on the strategy of the fight, and you will not feel the other fellow's punches. Once, when I thought I was about to drown in a seaplane accident, the freezing water seemed to me tepid. Or, more exactly, my consciousness was not concerned with the temperature of the water. It was absorbed by other thoughts. The temperature of the water has left no trace in my memory. In the same way, Sagon's consciousness was filled to the brim with the problem of getting away from the plane. His universe was limited successively to the fate of his crew, the handle that governed the sliding hatch, the rip cord of the parachute.

The inter-com seemed to be working. "Are you there?" he had called out.

No answer.

"Nobody on board?" he had asked again.

No answer.

They must have jumped, Sagon had decided. And as he was sorry about those flames (his hands and face were already burnt), he had got out of his seat, climbed out on the fuselage, and crawled forward along the surface of the wing.

"I peered in. I couldn't see the observer."

The observer, killed instantly by the German fighters, had slumped down out of sight.

"Then I backed up and looked for the gunner. I couldn't see him, either."

But the same thing had happened to the gunner.

"I thought they must have jumped."

Once again Sagon turned the matter over in his mind.

"If I had known, I could have crawled back into the cockpit. The flames were not so high. I lay there on the wing, I don't know how long. I had stabilized the plane at an angle before crawling out. The going was smooth, the wind was bearable, and I felt fairly comfortable. I must have been out on that wing for some time. I didn't know what to do."

Not that Sagon had been faced with insoluble problems. He thought himself alone on board. The plane was burning. The fighters were still after it and spattering it with bullets. What Sagon was telling us was that he had felt no desire of any kind. He had felt nothing. He had time on his hands. He was floating in a sort of infinite leisure. And point by point I recognized the extraordinary sensation that now and then accompanies the imminence of death—a feeling of unexpected leisure, absolutely the contrary of the picture-book notion of breathless haste. Sagon had lain there on his wing, a creature flung out of the dimension of time.

"And then," he said, "I jumped. I made a bad job of it. I could feel myself twisting in the air and hesitated to pull the cord, thinking I might get tangled up in the 'chute. I waited until I had straightened out. I waited quite a long time."

What Sagon really remembered of his whole mishap, from beginning to end, was waiting. Waiting for the flames to rise higher.

Then waiting on the wing for Heaven knows what. And finally, falling freely through the air, still waiting.

This was Sagon himself who was doing these things—actually a Sagon more rudimentary, more simple than the Sagon I know: a Sagon a little perplexed, bored and slightly impatient as he felt himself drop into an abyss.

# VIII

We had been living for two hours at the centre of an external pressure reduced to two thirds of normal. The crew were being gradually used up. We exchanged hardly a word. Once or twice, very cautiously, I tried to work my rudder. I was not obstinate about it. Each time the same sensation, the same feeling of a gentle exhaustion, had come over me.

Dutertre, at work with his camera, was careful to let me know in plenty of time when his photography required that I bank. I would do the best I could with such control of the wheel as was still left to me. I would tilt the plane and pull towards me; and in a dozen or twenty separate efforts I would set her where Dutertre wanted her.

"Altitude?"

"Thirty-three thousand seven."

I was still thinking of Sagon. Man is always himself. In myself I have never met another than myself. Sagon knew only Sagon. He who dies, dies as he was. In the death of an ordinary miner, it is an ordinary miner who dies. Where is it to be found—that haggard dementia that writers have invented to fascinate us with?

I saw once in Spain a man hauled up, after several days of excavation, out of the cellar of a house that had been destroyed by a bomb. He was blinking, for the daylight hurt his eyes; and men were holding him up, for he was tottering.

A crowd stood round him in silence and with what seemed to me a sudden timidity. This man, resuscitated almost from the beyond, still covered in the rubble in which he had been buried, half stupefied by suffocation and hunger, was like some dim monster. When some one grew bold enough to ask him questions, and to the questions he lent a kind of pallid attention, the timidity of the crowd changed to uneasiness.

Those round him tried to unlock his secret with bungling keys—for who is there can formulate the right question? They asked him what he had felt, what he had thought of, what he had done in that grave. They flung bridges at random across an abyss, like men seeking to reach the night of the mind of one blind and deaf and dumb, and bring him help. But when, finally, he was able to answer, what he said was, "Yes, I heard a long tearing sound." Or he said, "I was terribly worried. I was down there a long time. I thought it would never end." Or, "My back hurt. It hurt pretty badly." It was a decent fellow talking only about a decent fellow.

"I was worried about my watch," he said. "It was a wedding present. I couldn't get my hand into my pocket. I wondered if the cave-in had . . ."

It goes without saying that life had taught this man suffering and impatience, taught him the love of familiar things. He had made use of the man he was to take account of his universe, though it were the universe of a cave-in in the night. And the fundamental question, the question nobody thought of asking him but which governed all their blundering questions—"Who were you? Who surged up in you?"—this question he would have been unable to answer before time had allowed him little by little to build up the legend of himself. He would have been able to answer only—"Why, me . . . myself."

No single event can awaken within us a stranger totally unknown to us. To live is to be slowly born. It would be a bit too easy if we could go about borrowing ready-made souls.

It is true that a sudden illumination may now and then light up a destiny and impel a man in a new direction. But illumination is vision, suddenly granted the spirit, at the end of a long and gradual preparation. Bit by bit I learnt my grammar. I was taught my syntax. My sentiments were awakened. And now suddenly a poem strikes me in the heart.

Piloting now my plane, I feel no love; but if this evening something is revealed to me, it will be because I shall have carried my heavy stones toward the building of the invisible structure. I am preparing a celebration. I shall not have the right to speak of the sudden apparition in me of another than myself, since it is I who am struggling to awaken that other within me.

There is nothing that I may expect of the hazard of war except this slow apprenticeship. Like grammar, it will repay me later.

For us in the plane, life was losing its edge, blunted by a slow wearing away of ourselves. We were aging. The sortie was aging. What price high altitude? An hour of life spent at thirty-three thousand feet is equivalent to what? To a week? three weeks? a month of organic life, of the work of the heart, the lungs, the arteries? Not that it signifies. My semi-swoonings have added centuries to me: I float in the serenity of old age.

How far away now is the agitation in which I dressed! In what a distant past it is lost! And Arras is infinitely far in the future. The adventure of war? Where is there adventure in war? I have this day taken an even chance to disappear, and I have nothing to report unless it is that passage of tiny wasps seen for three seconds. The real adventure would have lasted but the tenth of a second; and those among us who go through it do not come back, never come back, to tell the story.

"Give her a kick to starboard, Captain."

Dutertre has forgotten that my rudder is frozen. I was thinking of a picture that used to fascinate me when I was a child. Against the background of an aurora borealis it showed a graveyard of fantastic ships, motionless in the Antarctic seas. In the ashen glow of an eternal night the ships raised their crystallized arms. The atmosphere was of death, but they still spread sails that bore the impress of the wind as a bed bears the impress of a shoulder, and the sails were stiff and cracking.

Here too everything was frozen. My controls were frozen. My machine-guns were frozen. And when I had asked the gunner about his, the answer had come back, "Nothing doing, sir."

Into the exhaust pipe of my mask I spat icicles fine as needles. From time to time I had to crush the stopper of frost that continued to form inside the flexible rubber, lest it suffocate me. When I squeezed the tube I felt it grate in my palm.

"Gunner! Oxygen all right?"

"Yes, sir."

"What's the pressure in the bottles?"

"Er . . . seventy. Falling, sir."

Time itself had frozen for us. We were three old men with white beards. Nothing was in motion. Nothing was urgent. Nothing was cruel.

The adventure of war. Major Alias had thought it necessary to say to me one day, "Take it easy, now!"

Take what easy, Major Alias? The fighters come down on you

like lightning. Having spotted you from fifteen hundred feet above you, they take their time. They weave, they orient themselves, take careful aim. You know nothing of this. You are the mouse lying in the shadow of the bird of prey. The mouse fancies that it is alive. It goes on frisking in the wheat. But already it is the prisoner of the retina of the hawk, glued tighter to that retina than to any glue, for the hawk will never leave it now.

And thus you, continuing to pilot, to daydream, to scan the earth, have already been flung outside the dimension of time because of a tiny black dot on the retina of a man.

The nine planes of the German fighter group will drop like plummets in their own good time. They are in no hurry. At five hundred and fifty miles an hour they will fire their prodigious harpoon that never misses its prey. A bombing squadron possesses enough firing power to offer a chance for defense; but a reconnaissance crew, alone in the wide sky, has no chance against the seventy-two machine guns that first make themselves known to it by the luminous spray of their bullets. At the very instant when you first learn of its existence, the fighter, having spat forth its venom like a cobra, is already neutral and inaccessible, swaying to and fro overhead. Thus the cobra sways, sends forth its lightning, and resumes its rhythmical swaying.

Each machine-gun fires fourteen hundred bullets a minute. And when the fighter group has vanished, still nothing has changed. The faces themselves have not changed. They begin to change now that the sky is empty and peace has returned. The fighter has become a mere impartial onlooker when, from the severed carotid in the neck of the reconnaissance pilot, the first jets of blood spurt forth. When from the hood of the starboard engine the hesitant leak of the first tongue of flame rises out of the furnace fire. And the cobra has returned to its folds when the venom strikes the heart and the first muscle of the face twitches. The fighter group does not kill. It sows death. Death sprouts after it has passed.

Take what easy, Major Alias? When we flew over those fighters I had no decision to make. I might as well not have known they were there. If they had been overhead, I should never have known it.

Take what easy? The sky is empty.

The earth is empty.

Look down on the earth from thirty-three thousand feet, and man ceases to exist. Man's traces are not to be read at this distance. Our telescopic lenses serve here as microscopes. It wants this microscope—not to photograph man, since he escapes even the telescopic lens—to perceive the signs of his presence. Highways, canals, convoys, barges. Man fructifies the microscope slide. I am a glacial scientist, and their war has become for me a laboratory experiment.

"Are the anti-aircraft firing, Dutertre?"

"I believe they are firing, Captain."

Dutertre cannot tell. The bursts are too distant and the smoke is blended in with the ground. They cannot hope to bring us down by such vague firing. At thirty-three thousand feet we are virtually invulnerable. They are firing in order to gauge our position, and probably also to guide the fighter groups towards us. A fighter group diluted in the sky like invisible dust.

The German on the ground knows us by the pearly white scarf which every plane flying at high altitude trails behind like a bridal veil. The disturbance created by our meteoric flight crystallizes the watery vapor in the atmosphere. We unwind behind us a cirrus of icicles. If the atmospheric conditions are favorable to the formation of clouds, our wake will thicken bit by bit and become an evening cloud over the countryside.

The fighters are guided towards us by their radio, by the bursts on the ground, and by the ostentatious luxury of our white scarf. Nevertheless we swim in an emptiness almost interplanetary. Everything round us and within us is total immobility.

We are now flying at three hundred and twenty-five miles an hour, you on the ground would say. But that is a race-course point of view. Here time is not, but only space. The earth itself, despite its twenty-five miles a second, moves but slowly round the sun. A whole year goes to the task. Perhaps we too are slowly approached in this exercise in gravitation. The density of aerial warfare? Grains of dust in a cathedral. We, grains of dust, are perhaps attracting to ourselves some dozens, it may be hundreds, of enemy grains of dust. And all those cinders rise as from a shaken rug slowly into the sky.

Take what easy, Major Alias? Looking straight down, all that I see is the bric-a-brac of another age exhibited under a pure crystal without tremor. I am leaning over the glass cases of a museum. But already the exhibit stands outlined against the light. Very far ahead lie Dunkerque and the sea. To left and right I see nothing. The sun has dropped too low, now, and I command the view of a vast glittering sheet.

"Dutertre! Can you see anything at all in this mess?"

"Straight down, yes."

"Gunner! Any sign of the fighters?"

"No sign, sir."

The fact is, I have absolutely no idea whether or not we are being pursued, and whether from the ground they can or cannot see us trailed by the collection of gossamer threads we sport.

Gossamer threads set me daydreaming again. An image comes

into my mind which for the moment seems to me enchanting. ". . . As inaccessible as a woman of exceeding beauty, we follow our destiny, drawing slowly behind us our train of frozen stars."

"A little kick to port, Captain."

There you have reality. But I go back to my shoddy poetry: "We bank, and a whole sky of suitors banks in our wake."

Kick to port, indeed! Try it.

The woman of exceeding beauty has fumbled her bank.

Is it true that I was humming?

For Dutertre has spoken again. "Hum like that, Captain, and you'll pass out."

He has certainly killed my taste for humming.

"I've just about got all the photos I want, Captain. Another few minutes and we can make for Arras."

We can make for Arras. Why, of course. Since we're half way there, we might as well.

Phew! My throttles are frozen!

And I say to myself:

"This week, one crew out of three has got back. Therefore, there is great danger in this war. But if we are among those that get back, we shall have nothing to tell. I have had adventures—pioneering mail lines; being forced down among rebellious Arabs in the Sahara; flying the Andes. But war is not a true adventure. It is a mere ersatz. Where ties are established, where problems are set, where creation is stimulated—there you have adventure. But there is no adventure in heads-or-tails, in betting that the toss will come out life or death. War is not an adventure. It is a disease. It is like typhus."

Perhaps I shall feel later that my sole veritable adventure in this war was that of my room in Orconte.

# IX

Orconte is a village on the outskirts of Saint-Dizier where my Group was stationed during the bitterly cold winter of '39. I was billeted in a clay-walled peasant house. The temperature would drop during the night low enough to freeze the water in my rustic crock, and the first thing I did in the morning was of course to light a fire. But to do that I had to get out of a bed in which I lay snug and warm and happy.

Nothing seemed to me more miraculous than that simple bed in that bare and freezing chamber. It was there that I revelled in the bliss of relaxation after the exhaustion of the day's work. I felt safe in that bed. No danger could reach me there. During the day I was exposed to the rigor of the upper altitudes and the risk of the peremptory machine guns. During the day my body was available for transformation into a lair of agony and undeserved laceration. During the day my body was not mine. Was no longer mine. Any of its members might at any moment be commandeered; its blood might at any moment be drawn off without my acquiescence. For it is another consequence of war that the soldier's body becomes a stock of accessories that are no longer his property. The bailiff arrives and demands a pair of eyes—you yield up the gift of sight. The bailiff arrives and demands a pair of legs—you yield up the gift of movement. The bailiff arrives torch in hand and demands the flesh off your face—and you, having yielded up the gift of smiling and manifesting your friendship for your kind, become a monster. Thus this body, which during any daylight hour might reveal itself my enemy and do me ill, might transform itself into a generator of whimperings, was still my obedient and comradely friend as it snuggled under the eiderdown in its demi-slumber, murmuring to my consciousness no more than its gratification and its purring bliss. Yet this body had to be withdrawn from beneath that eiderdown; it had to be washed in freezing water, shaved, dressed, made respectable before presenting itself to the bursts of steel. And getting out of bed was like a return to infancy, like being torn away from the maternal arms, the maternal breast, from everything that cherishes, caresses, shelters the existence of the infant.

So, having pondered and meditated and put off my decision as long as I could, I would grit my teeth and spring in a single leap to the fireplace, drench the logs with kerosene, and touch a match to them. Then, when the oil had flared up, and I had succeeded in crossing back to my bed, I would snuggle down again in its grateful warmth. With blankets and eiderdown drawn up to my left eye, I would watch the fireplace. At first the logs would seem not to catch, and only occasional flashes would flicker on the ceiling. But soon the fire would settle down in the hearth as if to organize a celebration. There would come a crackling, a roaring, a singing, and the fire would be as merry as a village wedding feast when the guests have begun to drink, to warm up, to nudge one another in the ribs.

Now and then it would seem to me that my good-tempered fire was standing guard over me like a particularly brisk and faithful shepherd dog going diligently about his work. A feeling of quiet jubilation would go through me as I watched it. And when the merry-making was at its height, when the shadows were dancing on the ceiling, when the warm golden music filled the air and the glowing logs had become a rosy architecture; when my room was quite redolent of the magic odor of smoke and resin, I would leap again from one friend to the other, from my bed to my fire; and standing there beside the more generous friend, I could never say whether I was in truth toasting my belly or warming my heart at that fireplace. Faced by two temptations, I like a coward had given way to the stronger, the ruddier, the one which, with its fanfare and flutter, had advertised its wares more cleverly.

Thus three times—first to light my fire, then to get back into bed, then again to harvest my crop of flames—three times with chattering teeth I had crossed the bare and frozen tundra of my chamber and known what it was to explore the polar regions. I had made my way on foot across a desert to arrive at a blessed haven, and my effort had been rewarded by that fire which in my presence, for my sake, had danced its jubilant air.

Very likely my story seems to you pointless, and yet this was a great adventure. My chamber had shown me as in a glass something I should never have discovered had I happened in by chance on this peasant house. What, as tourist, I should have seen would have been a bare and commonplace room, a vague bed, a water pitcher, an ugly chimney-piece. I should have yawned and turned away. Of its three provinces, its three civilizations—the one of sleep, the other of fire, the third of desert—I should have known nothing, nor been able to distinguish between them. How should I possibly have guessed the adventure of the body—first as infant clinging to the tenderness and the shelter of the maternal breast, then as soldier made for suffering, and finally as man enriched by the delight of the civilization of fire—fire, the magnetic pole of the tribe, that honors me and will do honor to my comrades who, when they come to see me if I get back, will take part in this festivity, will draw up their chairs round mine, and while we talk of our problems, our worries, our drudgery, will nevertheless say as they rub their hands and stuff their pipes, "There's no getting round it, a fire does make you feel fine."

But here in this plane there is no fire to persuade me to believe in friendship. There is no freezing chamber here to persuade me of the existence of adventure. I waken out of my reverie. There is nothing here but a void. Nothing but extreme old age. Nothing but a voice—Dutertre's, stubborn in its chimerical longing—saying to me:

"Give her a little kick to starboard, Captain."

# X

I am doing my job like a conscientious workman. Which does not alter the fact that I feel myself to be a pilot of defeat. I feel drenched in defeat. Defeat oozes out of every pore, and in my hands I hold a pledge of it.

For my throttle controls are frozen. The cold has turned them into two stumps of useless metal and has involved me in a serious predicament. For, whatever happens, I am forced to go on flying full throttle. Meanwhile, the pitch of my propellers, which serves in a sense as a brake on the revolution of my engines, is limited by an automatic check. If for any reason I am forced to dive, I shall be unable to reduce the speed of my engines, and unable also to increase my pitch. As I fall through space the torrential rush of air through my propellers will very likely increase the rotation of my engines to the point at which they blow up.

I could, if I had to, switch off my engines; but in that case I should never be able to start them again. I should then be stalled for good and all, which would mean the failure of the sortie and the crack-up of the machine. Not every terrain is favorable to the landing of a plane at one hundred and twenty miles an hour—and this, by manœuvering and gliding, is about the minimum speed at which I could hope to set the machine down. Therefore I must succeed in unblocking my throttles.

I was able to unblock the throttle of the port engine: the starboard throttle would not budge.

Now if I were forced down, I could reduce the speed of the port engine. But if I cut down the port engine, over which I have regained control, I should need to be able to offset the lateral traction exercised by the starboard engine—for the accelerated rotation of the starboard engine would obviously tend to pivot the plane to port. There is a way of offsetting this tendency. I could do it by the play of my rudder. But the bar that governs my rudder has long been frozen stiff. Therefore I should be able to offset nothing at all. The moment I cut down my port engine I must go into a spin.

Here was another of the war's absurdities. Nothing worked properly. Our world was made up of gear-wheels that would not mesh. And where the gear-wheels will not mesh, there is obviously no watchmaker.

After nine months of war we had still not succeeded in persuading the industries concerned that aerial cannon and controls ought to be manufactured with regard to the climate of the upper altitudes in which they were employed. What we were up against was not the irresponsible attitude of the manufacturers. Men are for the most part decent and conscientious. I am sure that almost always their seeming lack of initiative is a result and not a cause of their ineffectualness.

Ineffectualness weighed us down, all of us in the uniform of France, like a sort of doom. It hung over the infantry that stood with fixed bayonets in the face of German tanks. It lay upon the air crews that fought one against ten. It infected those very men whose job it should have been to see that our guns and controls did not freeze and jam.

We were living in the blind belly of an administration. An administration is a machine. The more perfect the machine, the more human initiative is eliminated from it. If, into a perfect machine, you introduce steel at one end, automobiles will come out of the other end. There will be no room for technical flaws, errors of measurement, human carelessness. And in a perfect administration, where man plays the part of a cog, such things as laziness, dishonesty, or injustice, cannot prevail.

But a machine is not built for creation. It is built for administration. It administers the transformation of steel into motor cars. It goes unvaryingly through motions pre-ordained once and for always. And an administration, like a machine, does not create. It carries on. It applies a given penalty to a given breach of the rules, a given method to a given aim. An administration is not conceived for the purpose of solving fresh problems. If, into your automobile-manufacturing machine, you inserted wood at one end, furniture would not come out at the other end. For this to happen, a man would have to intervene with authority to rip the whole thing up. But an administration is conceived as a safeguard against disturbances resulting from human initiative. The gear-wheels of the watch stand guard against the intervention of man. The watchmaker has no place among them.

I was posted to Group 2-33 in November 1939. When I arrived, my fellow pilots gave me due warning.

"You'll be flying over Germany," they said, "without guns or controls."

And to console me, they added: "But don't take it too hard, for it really doesn't matter. The German fighters always down you before you know they are there."

Six months later, in May 1940, the guns and the controls were still freezing up.

In the spring of 1940, everybody was repeating an ancient French saw: "France is always saved at the eleventh hour by a miracle."

There was a reason for the miracle. It used to happen occasionally that the beautiful administrative machine would break down and everybody would agree that it could not be repaired. For want of better, men would be substituted for the machine. And men would save France.

If a bomb had reduced the Air Ministry to ashes, a corporal—any corporal at all—would have been summoned, and the government would have said to him:

"You are ordered to see that the controls are thawed out. You have full authority. It's up to you. But if they are still freezing up two weeks from now you go to prison."

The controls would perhaps have been thawed out.

I could cite a hundred examples of this flaw. The Requisitions Committee for the Department of the North, for example, used to requisition heifers quick with young, and the slaughter-houses of France were transformed into graveyards of fœtuses. The requisitioning administration was a perfect machine. And because it was, not a single cog in the machine, not a single colonel on the board, had the slightest authority to act otherwise than as a cog. Each cog, as if the machine were a watch, was obedient to another cog. Revolt against the whole was useless. And this is why, once the machine began to go out of order, the cogs light-heartedly took to slaughtering freshened heifers. It may have been the lesser evil. Had the machine broken down altogether, the cogs might have begun to slaughter colonels.

I sat at my wheel discouraged to the marrow of my bones by this universal dilapidation. But as it seemed to me useless to blow up one of my engines, I fought again with the starboard throttle. In my disgust I forgot myself, wrestled with it too strenuously, and had to give it up. The effort had cost me another twinge at the heart. It was obvious that man was not made to do physical culture exercises at thirty-three thousand feet in the air. That twinge of pain was a warning, a sort of localized consciousness queerly come to life in the night of my organs.

"Let the engines blow up if they want to," I said to myself, "I don't care a hang." I was trying to catch my breath. It seemed to me that if I took my mind off my breath I should never be able to catch it again. The image of a pair of old-fashioned bellows came into my mind. I am stirring up my fire, I thought. And I prayed that it would make up its mind to catch.

Was there something I had wrenched beyond repair? At thirty-three thousand feet a slightly strenuous physical effort can strain the heart muscles. A heart is a frail thing. It has to go on working a long time. It is silly to endanger it for such coarse work. As if one burnt up diamonds in order to bake a potato.

## XI

As if one burnt up all the villages of France without by their destruction halting the German advance for a single day. And yet this stock of villages, this heritage, these ancient churches, these old houses with all the cargo of memories they carry, with their shining floors of polished walnut, the white linen in their cupboards, the laces at their windows that have served unfrayed so many generations—here they are burning from Alsace to the sea.

Burning is a great word when you look down from thirty-three thousand feet; for over the villages and the forests there is nothing to be seen but a pall of motionless smoke, a sort of ghastly whitish jelly. Below it the fires are at work like a secret digestion. At thirty-three thousand feet time slows down, for there is no movement here. There are no crackling flames, no crashing beams, no spirals of black smoke. There is only that grayish milk curdled in the amber air. Will that forest recover? Will that village recover? Seen from this height, France is being undermined by the secret gnawing of bacteria.

About this, too, there is much to be said. "We shall not hesitate to sacrifice our villages." I have heard these words spoken. And it was necessary to speak them. When a war is on, a village ceases to be a cluster of traditions. The enemy who hold it have turned it into a nest of rats. Things no longer mean the same. Here are trees three hundred years old that shade the home of your family. But they obstruct the field of fire of a twenty-two-year-old lieutenant. Wherefore he sends up a squad of fifteen men to annihilate the work of time. In ten minutes he destroys three hundred years of patience and sunlight, three hundred years of the religion of the home and of betrothals in the shadows round the grounds. You say to him, "My trees!" but he does not hear you. He is right. He is fighting a war.

But how many villages have we seen burnt down only that war may be made to look like war? Burnt down exactly as trees are cut down, crews flung into the holocaust, infantry sent against tanks, merely to make war look like war. Small wonder that an unutterable disquiet hangs over the land. For nothing does any good.

One fact the enemy grasped and exploited—that men fill small space in the earth's immensity. A continuous wall of men along our front would require a hundred million soldiers. Necessarily, there were always gaps between the French units. In theory, these gaps are cancelled by the mobility of the units. Not, however, in the theory of the armored division, for which an almost unmotorized army is as good as unmanœuvrable. The gaps are real gaps. Whence this simple tactical rule: "An armored division should move against the enemy like water. It should bear lightly against the enemy's wall of defence and advance only at the point where it meets with no resistance." The tanks operate by this rule, bear against the wall, and never fail to break through. They move as they please for want of French tanks to set against them; and though the damage they do is superficial,—capture of unit Staffs, cutting of telephone cables, burning of villages,—the consequences of their raids are irreparable. In every region through which they make their lightning sweep, a French army, even though it seem to be virtually intact, has ceased to be an army. It has been transformed into clotted segments. It has, so to say, coagulated. The armored divisions play the part of a chemical agent precipitating a colloidal solution. Where once an organism existed they leave a mere sum of organs whose unity has been destroyed. Between the clots—however combative the clots may have remained—the enemy moves at will. An army, if it is to be effective, must be something other than a numerical sum of soldiers.

We stand to the enemy in the relation of one man to three. One plane to ten or twenty. After Dunkerque, one tank to one hundred. We have no time to meditate upon the past; no time to say to ourselves even this—that forty million farmers must lose an armament race run against eighty million industrial workers. We are engaged in the present. And the present is what it is. No sacrifice, at any moment, on any front, can serve to slow up the German advance.

Whence it comes that throughout the civil and military hierarchies, from the plumber to the minister of state, from the second-class private to the general, there reigns a sort of uneasiness which no one can or dares put into words. There is no dignity in sacrifice if it is mere parody or suicide. It is beautiful to sacrifice oneself. These die in order that the rest be saved. The flames are grimly fought when the conflagration has to be put out. Men fight to the death in the cut-off camp so that their rescuers may have time to come to their aid. Yes, but we are surrounded by the conflagration. We have no camp on which to fall back. We know no rescuers on whom we can pin our hope. And as for those for whom we fight, for whom we say we are fighting, what are we doing except, apparently, ensuring their murder? For the aeroplane, dropping its bombs on towns behind the lines, has made this such a war as was never dreamt of.

I was later to hear foreigners reproach France with the few bridges that were not blown up, the handful of villages we did not burn, the men who failed to die. But here on the scene, it is the contrary, it is exactly the contrary, that strikes me so powerfully. It is our desperate struggle against self-evident fact. We know that nothing can do any good, yet we blow up bridges nevertheless, in order to play the game. We burn down real villages, in order to play the game. It is in order to play the game that our men die.

Of course some are overlooked! Bridges are overlooked, villages are overlooked, men are allowed to continue alive. But the tragedy of this rout is that all its acts are without meaning. The soldier who blows up a bridge can only do it reluctantly. He slows down no enemy—he merely creates a ruined bridge. He destroys his country in order to turn it into a splendid caricature of war. But it was a real bridge, not a caricature, that was blown up.

If a man is to strive with all his heart, the significance of his striving must be unmistakable. The significance of the ashes of the village must be as telling as the significance of the village itself. But the ashes of our villages are meaningless. Our dead must be as meaningful as death itself. But our dead die in a charade. The enemy's hundred and sixty divisions are not impressed by our burnings and our dead.

The question used to be asked, Are our men dying well or badly? Meaningless question! The Staff know that a given town can hold out for three hours. Yet our men are ordered to hold it forever. Having no means of offense, they as good as beg the enemy to destroy the town in order that the rules of war be respected. They are like a friendly opponent at chess who says, "But you have forgotten to take your pawn." Our men spend their time challenging the enemy.

"We are the defenders of the village," they say in effect. "You are the attackers. Ready? Play!"

And under the burst of an enemy squadron the village is wiped out.

"Well played, Nazi!"

Certainly inert men exist, but inertia is frustrated despair. Certainly fugitives exist, and I remember that twice or three times Major Alias had threatened to shoot occasional gloomy wretches picked up on the highways and evasive in the answers they gave to his questions. One's impulse is so strong to make somebody responsible for disaster, and to believe that by putting him out of the way all can be saved. The fugitives are responsible for the rout, since there would be no rout if there were no fugitives. Therefore, flourish a gun and all is well.

As well bury the sick in order to eliminate sickness. Major Alias always ended by slipping his gun back into its holster. He could see very well that there was something awfully pompous about that gun, like a comic-opera saber. Alias knew perfectly well that those mournful fellows were an effect, not a cause of the disaster. He knew absolutely that they were the same men, exactly the same men, as those who, somewhere else in France, at that very moment, were accepting the fact that they must die. In two short weeks one hundred and fifty thousand of them accepted the fact that they must die. But some men are stubborn and insist upon a reason why they should die.

It is hard to find a reason.

Here is a runner engaged in the race of life against other runners of his own class. The starter fires, the runner springs forward —and he discovers that he has a ball and chain attached to his leg. He quits.

"This race doesn't count," he says.

"It does though, it does!" you protest.

What are you going to tell a man to make him put his heart into a race that is not a race? Alias knew what those fugitives were thinking. "This race doesn't count," was what they were thinking.

Alias put his gun back into the holster and tried to find a better argument.

There is but one better argument, but one logical argument, and I challenge anybody to find another. It is this: "Your death will have no effect at all. Defeat is inescapable. But it is proper that a defeat manifest itself by dead. There must be mourning. Your part is to play the dead."

"Very good, sir."

Alias did not despise the fugitives. He knew well enough that his argument always worked. He himself accepted the expectancy of death. All his crews accepted the expectancy of death. His argument, slightly disguised, never failed to work with us: "It's damned awkward. But the General Staff want it done. They very much want it done. . . . And that's that."

"Very good, sir."

Alias knew that we had accepted.

My very simple notion is that those who died served as bondsmen for the rest.

# XII

I have aged so much that all that I was is left behind me. I stare out through the great glittering plate of my windscreen. Below me are men. Infusoria wriggling under a microscope. Who can work up interest in a family of infusoria?

Were it not for this twinge of pain that seems to me a living thing, I could sink into drowsy rumination, like an aged tyrant. It is only ten minutes since I spoke of our crews as supernumeraries. Pure rhetoric and sickeningly false. When I saw the German fighters below, did my fancy speak of tender sighs? It spoke of poisonous wasps. That was reality. They were tiny, and they were obscene. It is hard to believe that I invented that disgusting literary image of a dress with a train. I couldn't have! For one thing, I have never seen the wake of my ship. Here in this cockpit, in which I fit like a pipe in its case, I can see nothing behind me. I see behind me through the eyes of my gunner. And then only if the inter-com is working. My gunner never called down to me, "Adoring suitors aft in the wake of our train!"

All this is mere juggling with words. Of course I should like to believe, I should like to fight, I should like to win. But try as a man will to pretend to believe, pretend to fight, pretend to win by setting fire to his own villages, it is hard to feel elation over pretense.

It is hard to exist. Man is a knot into which relationships are tied, and my ties serve me hardly at all.

What is this in me that has broken down? What is the secret of substitutions? Whence comes it that a gesture, a word, can give rise to endless ripples in a human destiny? Whence comes it that in other circumstances I should be overwhelmed by what seems to me now remote and abstract? Whence comes it that if I were Pasteur, the play of true infusoria would seem to me pathetic to the point where a slide under a microscope would represent something infinitely more vast than a virgin forest, and the watching of that slide would seem to me the most thrilling kind of adventure? Whence comes it that that black dot below, which is a house of men. . . .

But again a childhood memory returns to me.

When I was a small boy. . . . I speak of my early childhood, that is to say, of a vast region out of which all men emerge. Whence come I? I come from my childhood. I come from childhood as from a homeland. . . . When I was a small boy, then, I had a queer experience.

I must have been five or six years old. It was eight in the evening. At eight o'clock children ought to be in bed. Particularly in winter, when night has already fallen. For some reason I had been forgotten.

On the ground floor of our house in the country—which was big—there was a hall that seemed to me immense. It led into the warm room at the back in which we children were fed our supper. I had always been afraid of that hall, perhaps because of the feeble light of the lamp that hung in the middle of it and scarcely drew it forth from the darkness. A signal rather than a light. The hall was paneled high up, and the paneling creaked, which was another reason for my fear. And it was cold. Coming into it out of the warm and lamplit rooms that lined it was like coming into a cavern.

But that evening, seeing that I had been forgotten, I gave way to the demon of evil in me, reached up on tiptoe for the handle of our supper-room door, pushed the door softly in, and embarked upon my illicit exploration of the world.

The creaking of the paneling was the first warning I received of heavenly anger. I could see in the shadow the great reproving panels. Not daring to explore farther, I climbed up on a console table, and there, resting against the wall and letting my legs hang, I sat with beating heart like every shipwrecked sailor before me on his reef in mid-sea.

At that moment the drawing-room door opened. Two uncles who absolutely terrified me shut the door behind them upon the lights and the hubbub of voices, and began to pace the hall.

I trembled lest I be discovered. Uncle Hubert was in my eyes the very image of severity. A delegate of divine justice. This man, who never in his life had tweaked a child's ear or pinched its cheek affectionately, always threatened me when I had been naughty with a terrifying frown and these words: "The next time that I go to America I shall bring back a whipping machine. American machines are the most modern in the world. That is why American children are the best behaved in the world. And a very good thing for their parents, too."

I did not like America.

Here they were, then, strolling back and forth through the interminable hall while I almost fainted holding my breath and following them with my eyes and ears. "In times like these," they said; and they moved off with their secret meant only for grown people. "In times like these," I memorized the phrase. Then, as if a tide had rolled up to me another of its indecipherable treasures—"It's pure madness, positive madness," one uncle said to the other. And I fished up that phrase as if it were a priceless thing, and to myself I said slowly, testing its power upon the consciousness of a five-year-old, "It's pure madness, positive madness."

The tide carried my uncles away, the tide rolled them up again. With a kind of sidereal regularity, like a gravitational phenomenon, this going and coming repeated itself and suggested to me fitfully lighted glimpses of the life of man. I was marooned on my console for eternity, the clandestine listener to a solemn consultation in the course of which my uncles, who knew all there was to know, were collaborating in the creation of the world. The house might stand a thousand years: for all that thousand years my two uncles, pacing the hall with the patience of a pendulum, would continue to fill the air with the apprehension of eternity.

That black dot at which I stare is surely a human habitation thirty-three thousand feet below me. And I receive nothing from it. Yet it is possibly a great country house, and there may be two uncles in it pacing to and fro and slowly constructing in the consciousness of a child something as fabulous as the immensity of the seas.

My field of vision embraces a territory as large as a province, yet round me space has shrunk to the point of suffocation. In all this space I have less space at my disposal than was available to me in the replica of that black dot. I have lost the sense of distance, am blind to distance. But I feel now a kind of thirst for it. And it seems to me that I have stumbled here upon a common denominator of all the aspirations of mankind.

When chance awakens love, everything takes it place in a man in obedience to that love, and love brings him the sense of distance. When, in the Sahara, the Arabs would surge up in the night round our campfires and warn us of a coming danger, the desert would spring to life for us and take on meaning. Those messengers had lent it distance. Music does something like this. The humble odor of an old cupboard does it when it awakens and brings memories to life. Pathos is the sense of distance.

But I know that nothing which truly concerns man is calculable, weighable, measurable. True distance is not the concern of the eye; it is granted only to the spirit. Its value is the value of language, for it is language which binds things together.

And now it seems to me that I begin to see what a civilization is. A civilization is a heritage of beliefs, customs, and knowledge slowly accumulated in the course of centuries, elements difficult at times to justify by logic, but justifying themselves as paths when they lead somewhere, since they open up for man his inner distance.

There is a cheap literature that speaks to us of the need of escape. It is true that when we travel we are in search of distance. But distance is not to be found. It melts away. And escape has never led anywhere. The moment a man finds that he must play the races, go to the Arctic, or make war in order to feel himself alive, that man has begun to spin the strands that bind him to other men and to the world. But what wretched strands! A civilization that is really strong fills man to the brim, though he never stir. What are we worth when motionless, is the question.

There is a density of being in a Dominican at prayer. He is never so much alive as when prostrate and motionless before his God. In Pasteur, holding his breath over the microscope, there is a density of being. Pasteur is never more alive than in that moment of scrutiny. At that moment he is moving forward. He is hurrying. He is advancing in seven-league boots, exploring distance despite his immobility. Cézanne, mute and motionless before his sketch, is an inestimable presence. He is never more alive than when silent, when feeling and pondering. At that moment his canvas becomes for him something wider than the seas.

Distance granted man by the childhood home, by the chamber at Orconte, by the field of vision of Pasteur's microscope; distance opened up by a poem. What are these but the fragile and magical gifts that only a civilization is able to distribute? For distance is the property of the spirit, not of the eye; and there is no distance without language.

But how am I to quicken the sense of my language when all is confusion? When the trees round the house are at one and the same time a ship transporting the generations of a family and a mere screen in the way of an artilleryman? When the press of the German bombers bearing down upon the villages has squeezed out a whole people and sent it flowing down the highways like a black syrup? When France displays the sordid disorder of a scattered ant-hill? When we must fight, not against a flesh-and-blood opponent, but against rudders that freeze, throttles that jam, bolts that stick?

"You may drop down now, Captain."

I may drop down. I shall drop down. I shall drop down upon Arras. I shall carry out the second half of our mission—the low-altitude sortie. Behind me I have a thousand years of civilization to help me. But they have not helped me yet. I dare say this is not the moment for rewards.

At five hundred miles an hour I lose altitude. Banking, I have left behind me a polar sun exaggeratedly red. Ahead and three or four miles below me, I see the broad surface of a rectilinear mass of cloud that looks like an ice-floe. A whole province of France lies buried in its shadow. Arras lies shadowed by it. Beneath my ice-floe, I imagine, the world has a blackish tinge. The war must be stewing there as in the belly of a giant soup-kettle. Jammed roads, flaming houses, tools lying where they were flung down, villages in ruins, muddle, endless muddle.

To drop down here is like tumbling into a ruin. We shall have to splash about in their mud. We shall have to live with those below in their barbarous dilapidation. Below us lies a world in decomposition. We are like travelers who, after long years amid coral and palm, are on our way home penniless. We face the prospect of a return to our native sordidness—the greasy food of avaricious relatives, the cantankerousness of family squabbles, the bad conscience born of money cares, the disappointed hopes, the degrading flight before the rent-collector, the arrogance of the landlord; squalor, and the stinking death in hospital. Up here at any rate death is clean. A death of flame and ice. Of sun and sky and flame and ice. But below! That digestion stewing in slime . . .

# XIII

Dutertre's voice came down the inter-com.

"Due south, Captain."

Quite right. Safer to lose altitude over our own zone than the enemy's.

Looking down on those swarming highways I understood more clearly than ever what peace meant. In time of peace the world is self-contained. The villagers come home at dusk from their fields. The grain is stored up in the barns. The folded linen is piled up in the cupboards. In time of peace each thing is in its place, easily found. Each friend is where he belongs, easily reached. All men know where they will sleep when night comes. Ah, but peace dies when the framework is ripped apart. When there is no longer a place that is yours in the world. When you know no longer where your friend is to be found. Peace is present when man can see the face that is composed of things that have meaning and are in their place. Peace is present when things form part of a whole greater than their sum, as the divers minerals in the ground collect to become the tree.

But this is war.

I can see from my plane the long swarming highways, that interminable syrup flowing endless to the horizon. The inhabitants of the war zone are being evacuated. This, at any rate, is the official version. But it is no longer true. They are evacuating themselves. There is a crazy contagion in this exodus. Where are these vagabonds going? They are going south—as if in the south there was room for them, food for them, tender hands waiting to welcome them. But southward there are only villages filled to bursting, men and women sleeping in sheds, stocks of food running out. Southward the most generous hearts are beginning little by little to harden at the sight of this mad invasion which little by little, like a sluggish river of mud, is beginning to suffocate them. Can a single province lodge and nourish all France?

Where are they going? They have no notion. They are tramping towards phantom havens—for scarcely does this caravan come up to an oasis when it ceases to be an oasis. One by one each oasis bursts its bonds and pours into the caravan. And when, by chance, the caravan comes upon a real village, a village that seems still to be alive, it swallows up its substance in a single night, gnaws it clean as the worm polishes the bone.

Faster than the exodus, the enemy moves. Here and there armored cars roll past the stream. It thickens, swirls, flows for a moment backwards. Whole German divisions flounder in this stew; and Germans who at another point were killing their kind are here quenching the thirst of the refugees.

In the course of the retreat our Group had been quartered in a dozen villages. A dozen times our Group had been entangled in the dragging herd that shuffled slowly through those villages.

"Where are you bound?"

"Nobody knows."

They never knew. Nobody knew anything. They were evacuating. There was no way to house them. Every road was blocked. And still they were evacuating. Somewhere in the north of France a boot had scattered an ant-hill, and the ants were on the march. Laboriously. Without panic. Without hope. Without despair. On the march as if in duty bound.

"Who ordered you to evacuate?"

It was always the mayor, or the schoolteacher, or the mayor's clerk. One morning at three the order had run through the village:

"Everybody out!"

They had been expecting this. For two weeks they had seen the passage through their village of refugees who no longer believed in the eternity of their homes. Man had been a settler on the planet. He had ceased to be a nomad. He had built himself villages that had lasted through the ages. He had waxed and polished floors and chairs that had gone on serving his great-grandchildren. The family house had received him at his birth, transported him to his death; and then, like a good bark crossing the water from bank to bank, it had carried his sons over the same stream. All that was ended now. The villagers were on the move. And no one so much as knew why.

The highways too were part of our experience. We were pilots, and there were days when in a single morning our sortie took us over Alsace, Belgium, Holland, and the sea itself. But our problems were most often of the north of France, and our horizon was very often limited to the dimensions of a traffic tangle at a crossroads. Thus, only three days earlier, I had seen the village in which we were billeted go to pieces. I do not expect ever to be free of that clinging, viscous memory.

It was six in the morning, and Dutertre and I, coming out of our billet, found ourselves in the midst of chaos. All the stables, all the sheds, all the barns and garages had vomited into the narrow streets a most extraordinary collection of contrivances. There were new motorcars, and there were ancient farm carts that for half a century had stood untouched under layers of dust. There were hay wains and lorries, carryalls and tumbrils. Had we seen a mail-coach in this maze it would not have astonished us. Every box on wheels had been dug up and was now laden with the treasures of the home. From door to vehicle, wrapped in bedsheets sagging with hernias, the treasures were being piled in.

Together, these treasures had made up that greater treasure—a home. By itself, each was valueless; yet they were the objects of a private religion, a family's worship. Each filling its place, they had been made indispensable by habit and beautiful by memory, had been lent price by the sort of fatherland which, together, they constituted. But those who owned them thought each precious in itself and for itself. These treasures had been wrenched from their fireside, their table, their wall; and now that they were heaped up in disorder, they showed themselves to be the worn and torn stock of a junk-shop that they were. Fling sacred relics into a heap, and they can turn your stomach.

"What's going on here? Are you mad?"

The café owner's wife shrugged her shoulders.

"We're evacuating."

"But why, in God's name!"

"Nobody knows. Mayor's orders."

She was too busy to talk, and vanished up her staircase. Dutertre and I stood in the doorway and looked on. Every motorcar, every lorry, every cart and charabanc was piled high with children, mattresses, kitchen utensils.

Of all these objects the most pitiful were the old motorcars. A horse standing upright in the shafts of a farm-cart gives off a sensation of solidity. A horse does not call for spare parts. A farm-cart can be put into shape with three nails. But all these vestiges of the mechanical age! This assemblage of pistons, valves, magnetos, and gear-wheels! How long would it run before it broke down?

"Please, Captain. Could you give me a hand?"

"Of course. What is it?"

"I want to get my car out of the garage."

I looked at the woman in amazement.

"Are you sure you know how to drive?"

"Oh, it will be all right. The road is so jammed, it won't be hard."

There was herself, and her sister-in-law, and their children—seven children in all.

That road easy to drive? A road over which you made two or ten miles a day, stopping dead every two hundred yards? Braking, stopping, shifting gears, changing from low to second and back again every fifty yards in the confusion of an inextricable jam. Easy driving? The woman would break down before she had gone half a mile! And gas! And oil! And water, which she was sure to forget!

"Better watch your water. Your radiator is leaking like a sieve."

"Well, it's not a new car."

"You'll be on the road a week, you know. How are you going to make it?"

"I don't know."

She won't have gone three miles before running into half a dozen cars, stripping her gears, and blowing out her tires. Then she and her sister-in-law and the seven children will start to cry. And she and her sister-in-law and the seven children, faced by problems out of their ken, will give up. They will abandon the car, sit down by the side of the road, and wait for the coming of a shepherd.

But it is astonishing how few shepherds there are. Dutertre and I are staring at sheep who have taken things into their own hands. And these sheep are off in an immense clatter of mechanical equipment. Three thousand pistons. Six thousand valves. The grate, the grind, the clank of this machinery. Water boiling up in a radiator already. And slowly, laboriously, this caravan of doom stirs into movement. This caravan without spare parts, without tires, without gasoline, without a mechanic. They are mad!

"Why don't you stay home?"

"God knows, we'd rather stay."

"Then why do you leave?"

"They said we had to."

"Who said so?"

"The mayor."

Always the mayor.

"Of course we'd all rather stay home."

It is a fact that these people are not panicky; they are people doing a blind chore. Dutertre and I tried to shake some of them out of it.

"Look here, why don't you unload and put that stuff back into your house. At least you'll have your pump-water to drink."

"Of course that would be the best thing."

"But you are free to do it. Why don't you?"

Dutertre and I are winning. A cluster of villagers has collected round us. They listen to us. They nod their heads approvingly.

"He's right, he is, the captain."

Others come to our support. A roadmender, converted, is hotter about it than I am.

"Always said so. Get out on that road and there's nothing but asphalt to eat."

They argue. They agree. They will stay. Some go off to preach to others. And they come back discouraged.

"Won't do. Have to go."

"Why?"

"Baker's already left. Who will bake our bread?"

The village has already broken down. At one point or another it has burst; and through that hole its contents are running out. Hopeless.

Dutertre said what he thought about it:

"The tragedy is that men have been taught that war is an abnormal condition. In the past they would have stayed home. War and life were the same thing."

The café owner came down, dragging a sack.

"You can let us have a cup of coffee, I suppose. We are flying in half an hour."

"Ah, my poor lads!"

She wiped her eyes. It was not us she was weeping for. Nor herself. Already she was crying with exhaustion. Already she felt herself suffocating in that caravan which was to go further to pieces with every mile of its journey.

Farther on, in the open country, the enemy fighters would be flying low and spitting forth their bursts of machine-gun fire upon this lamentable flock. But it was astonishing how on the whole the enemy refrained from total annihilation. Here and there stood a car in flames, but very few. And there were few dead. Death was a sort of luxury, something like a bit of advice. It was the nip in the hock by which the shepherd dog hurried the flock along. Though one wondered why the enemy action was so little insistent, so altogether sporadic and local. The enemy was at no pains whatever to blow the caravan to bits. True, the caravan had no need of the enemy to go to pieces. The machines took care of that. They went spontaneously out of order. The machine is conceived for a deliberate and peaceful society, a society master of its time. When man is not present to repair the machine, regulate it, polish it, it ages at a dizzying pace. Tonight all these machines will look a thousand years old. I seemed to be looking on at the death-throes of the machine.

Here is a peasant whipping up his horse. Perched on his seat with the majesty of a king, he lords it over the whole caravan.

"You look very satisfied up there."

"Ah, it's the end of the world."

Suddenly I felt queasy. All these workers, these simple people, each with his place in the world, were to be transformed into parasites, vermin. They were going to spread over the countryside and devour it. The thought made me sick.

"Who is going to feed you?"

"Nobody knows."

How is one to feed millions of migrants shuffling over miles of road at the rate of two to ten miles a day? If food existed, it could not be brought up to them.

All this muddle of men and old iron lost on the asphalt of the highways made me think suddenly of my march through the Libyan desert. Prévot and I had crashed in a landscape glassy with black rocks and covered with a carpet of sun-grilled iron. This was not far different.

I stared at the refugees in despair. How long would a swarm of locusts last in a field of asphalt?

"Do you expect to drink rain-water?"

"Nobody knows."

They knew nothing. For ten days they had seen an unbroken stream of refugees from the north flow through their village. For ten days they had watched this unending exodus. And their turn had come. They would take their place in the procession. But without confidence:

"If it was up to me, I'd rather die at home."

"We'd all rather die at home."

That was true. Their village might have collapsed over their heads, and still none would have chosen to leave.

Had France possessed reserves of food, that food could never have been brought up the highways down which this stream was flowing. If you have to, you can force your way downstream through brokendown cars, jammed cars, inextricable snarls of traffic at successive crossroads. But how can you move against such a stream?

"There being no reserves of food," said Dutertre grimly, "all is well."

A rumor is spreading that the Government has forbidden all evacuations. Even if it were true, how were the orders to be transmitted? There are no open roads, and the telephone cables are jammed, or cut; or the messages are received with a distrust born of experience. And it is no longer a matter of giving orders. What is wanted is the invention of a new code. For a thousand years man has been taught that women and children are to be shielded from war. War is a matter for men only. The village mayors are full of this law of society; their clerks know it; the schoolteachers know it. Assume that suddenly they receive orders to stop the evacuations, which is to say, force women and children to remain in the zone of bombardment. It will take them a month to adjust their conscience to this sign of a new age. You cannot overthrow a system of morality at one blow. And while you examine your conscience, the enemy continues his advance. Wherefore the mayors, their clerks, the schoolteachers send forth this stream of people on the highways. What is to be done? Where does truth reside? Forward troop the sheep without shepherd.

"Is there a doctor in this village?"
"You don't live here, I take it?"
"No. We live up north."
"What do you want of a doctor?"

"My wife is going to have a baby."

Lying among her kitchen utensils, in this desert of old iron.

"Couldn't you have thought of a doctor earlier?"

"We've been four days on the road."

The road is an irresistible stream. Where can you stop? Every village you move through is deserted the moment you arrive, pours into the caravan like the flow of a burst pipe into a giant sewer.

"No. No doctor here. The Group doctor is ten miles up the line."

"Well. Thank you."

The man mopped his forehead. Everything was going to pieces. His wife would bring her child into the world in a bed of kitchen utensils. There was nothing cruel about this. It was above all, most of all, monstrously beyond the bounds of things human. Nobody complained. Complaint was meaningless. His wife would die, and he would not complain. His wife was to die in childbed. Complain of what? There was no help for it. It was a nightmare.

"If we could only stop somewhere!"

Find a real village, a real inn, a real hospital. But, for God knows what reason, the hospitals too are being evacuated. It is part of the game. There isn't time to recast the rules of the game. Find a real death. But there is no real death any longer. There are bodies that break down the way the cars do.

Everywhere in this mob I sense a wearied haste, a haste that has renounced haste. At the rate of two to ten miles a day these people are fleeing before tanks moving at fifty miles a day and aeroplanes flying at four hundred miles an hour. Thus treacle flows when the bottle has been overturned. This man's wife would lie in; but he had all the time in the world before him. It was urgent. Was it really urgent? It was suspended in unstable equilibrium between urgency and eternity.

The world of these people had slowed down, like the reflexes of a dying man. This was an enormous flock that stood, exhausted and shuffling, at the gates of a slaughter-house. Were there ten or only five million of them on the asphalt? Here was a people accepting the notion of its reabsorption into eternity.

"How," I said to myself, "are these people to survive? Man does not eat branches." But they themselves were not in the least horrified by their fate. Wrenched from their homes, their work, their responsibilities, they had lost all significance. Their very identity seemed to have been rubbed off. They were very little themselves. They were very little alive. Later, they would re-invent their sufferings. Meanwhile they were suffering most of all from the aching strain of heavy loads, from the loosened knots in bed-sheets that dripped with their dreary entrails, from the strain of pushing motorcars forward in the attempt to make the engines turn over.

Not a word about defeat. Naturally. No man feels the need of discussing a thing which constitutes his very substance. They *were* the defeat. I had suddenly the vision of a France losing its entrails. Quick! Sew up our France! There is not a moment to lose! France is doomed.

It began again. Like fish on dry land, these people were suffocating:

"Anybody got any milk here?"

A question to make you die laughing.

"My kid hasn't drunk anything since yesterday."

The kid was a six-months-old baby. He made a lot of noise. But his noise wouldn't last. Fish out of water are soon quiet. There is no milk here. There is only scrap-iron here. There is only an enormous quantity of useless scrap-iron, falling apart mile after mile, dropping bolts, nuts, screws, sheets, while it bears this prodigiously needless exodus, this people, away towards oblivion.

A rumor spreads that some miles to the south the road is being machine-gunned by the enemy. There is talk of bombs. There is even the muffled sound of distant explosions. The rumor is no mere rumor.

But these people are not frightened. They seem even to perk up a little at the news. That concrete risk seems to them healthier than this drowning in old iron.

Ah, the blueprint that historians will draft of all this! The angles they will plot to lend shape to this mess! They will take the word of a cabinet minister, the decision of a general, the discussion of a committee, and out of that parade of ghosts they will build historic conversations in which they will discern farsighted views and weighty responsibilities. They will invent agreements, resistances, attitudinous pleas, cowardices. But I know what an evacuated ministry can be. I've seen one. It taught me that once a government evacuates, it is no longer a government. It is like a human body. If you begin to take it apart, sending the stomach here, the liver there, the guts somewhere else, that collection no longer constitutes an organism. I spent twenty minutes at the Air Ministry. And I can tell you that in time of evacuation a minister is a being who controls the movements of his messenger. Miraculous control. He has only to press a button. An electric cord still joins the flunkey to the minister. The minister presses the button and the flunkey appears.

"My car," says the minister.

And there his authority stops. He gives his flunkey a little exercise. But the flunkey is not sure if, on earth, there exists a car that is the minister's. No electric cord runs between the flunkey and anything at all. The chauffeur is lost somewhere, out in the world. What could the men who governed us know of the war? Situated as we were, impossible as liaison now was, it would take our people a week to arrange for the bombardment of an enemy division spotted by my Group. What sound could reach the ears of our governors from this land that was losing its entrails? News moved at the rate of ten miles a day. The telephone service was out. There was no way of transmitting a picture of this being, this France, in a state of decomposition. The Government swam in a void, a polar void. From time to time it was reached by desperately urgent appeals; but they were abstract, reduced to three scrawled lines. How could those who governed us know whether ten million Frenchmen had or had not already died of hunger? And this cry for help from ten million men could have been contained in a single sentence. It wants but a single sentence to say:

"Meet you tomorrow at four."

Or:

"They say ten million men are dead."

Or:

"Blois is in flames."

Or:

"They've found your chauffeur."

All this on the same level of importance. Just like that. Ten million men. The motorcar. The Army of the East. Western civilization. The chauffeur has been found. England. Bread. What time is it?

I give you seven letters. They are the seven letters of the Bible. Reproduce the Bible with them for me.

Historians will forget reality. They will invent thinking men, joined by mysterious fibers to an intelligible universe, possessed of sound farsighted views and pondering grave decisions according to the purest laws of Cartesian logic. There will be powers of good and powers of evil. Heroes and traitors. But treason implies responsibility for something, control over something, influence upon something, knowledge of something. Treason in our time is a proof of genius. Why, I want to know, are not traitors decorated?

# XIV

Already as I move in the direction of Arras, peace is everywhere beginning to take shape. Not that well-defined peace which, like a new period in history, follows upon a war decorously terminated by a treaty. This is a nameless peace that stands for the end of everything. For an end of things that go on endlessly ending. It is an impulse that little by little finds itself bogged down. There is no feeling that either a good or a bad conclusion is on the way. Quite the contrary. Little by little the notion that this putrefaction is provisional gives way to the feeling that it may be eternal. Nothing here is conclusive for there is no grip by which this great creature can be seized as you might seize a drowning man by knotting your fist in his hair. Everything has gone to pieces, and not even the most pathetic striving can bring up more than an insufficient lock of hair. The peace that is on its way is not the fruit of a decision reached by man. It spreads apace like a gray leprosy.

Below me, those roads on which the caravan is breaking down, on which the enemy kills or quenches thirst at will, put me in mind of the miry regions where land and water are indistinguishable. This peace, that has become fused with this war, has begun to rot this war.

My friend Léon Werth heard on the road an extraordinary remark which he recorded in his excellent book. On the left were the Germans, and on the right the French. Between the two flowed the sluggish stream of refugees. Hundreds of women and children extricating themselves as well as they were able from flaming motorcars. A French artillery officer, entangled despite himself in the snarl, stopped to set up his Seventy-five beside the road. Opposite, the enemy aimed, missed his target, and mowed down the migrants in his line of fire. Whereupon Frenchwomen rushed upon the French lieutenant who, running with sweat, was stubbornly performing his incomprehensible duty, trying (with twelve men!) to hold a position that was untenable, and shouted at him:

"Go away! Go away! You cowards!"

The lieutenant and his men went away. Wherever they went, they were brought up against problems of peace, not of war. Of course children should not be massacred on the highways. Yet every soldier who pulled a trigger found a child in his line of fire. Every French lorry that moved or tried to move through that mob was potentially the cause of death among those people. For, moving upstream against the flow, the lorry could not but bottle up the whole highway.

"You are mad! Let us through! The children are dying!"

"We're fighting a war."

"What war? Where are you fighting it? It will take you three days to go a mile against this current."

Here was a handful of French soldiers in a lorry, trying to reach a point to which they had been ordered and which had certainly been abandoned to the enemy hours before. All that they could think about was their plain duty:

"Gangway, there!"

"Why don't you let us ride with you? You are beasts!"

A child bawls.

"And the kid. . . ."

But the kid has stopped crying. It takes milk to make a child cry.

"We're fighting a war."

There was a kind of despairing stupidity in the way they repeated it.

"But you will never find your war! You will croak on the road with the rest of us!"

"We're fighting a war."

They were by no means sure of what they were saying. They were by no means sure that they were fighting a war. They had never seen the enemy. They were rolling in a lorry towards a goal more fugitive than a mirage. They were moving towards nothing more than a peace that was a pool of putrefaction. And as they were caught up inextricably in the chaos, they jumped down from the lorry. Instantly they were surrounded.

"Have you any water?"

So they shared their water.

"Have you any bread?"

And they shared their bread.

"But you can't leave her here to die!"

So into their lorry they put the woman who lay dying in a car wrecked by the side of the road.

"And what about this child?"

The child went in beside the dying woman.

"And this woman in labor."

They put her in beside the living child.

And for another woman they found room merely because she was crying so bitterly.

It took an hour to free the lorry and turn it round till it too faced south. Rising like an erratic block, it too would now be carried downstream by the civilian flood. The soldiers had been converted to this peace. Because they hadn't been able to find the war. Because the musculature of the war was invisible. Because the gun aimed at you kills a child. Because on your way up to the lines you stumble upon women in labor. Because it is as useless to try to transmit information or receive a command as to communicate with the inhabitants of Mars. There is no longer an army. There are only men.

They have been converted to this peace. They have been changed by the force of things into mechanics, doctors, shepherds, stretcher-bearers. Because, since these little people are ignorant of how to cure the ills of their scrap-iron, the soldiers have taken to repairing their cars. And not one of them could tell you, in the midst of his sweating labor, whether he was a hero or a man who deserved to be court-martialled. It would not astonish him if he were decorated on the spot. Nor if he were stood up against a wall with a dozen bullets in his skull. Nor if he were demobilized. Nothing would astonish him. It is a long time since he and his kind have crossed the frontiers of astonishment.

Here is an immense stew in which not an order, not a movement, not a scrap of news, not a wave of anything at all can run on beyond a single mile. Exactly as the villages topple one by one into the common sewer, so these army lorries, absorbed into this peace, are one by one converted to this peace. These handfuls of men who would have accepted without question the notion of their imminent death—assuming they had so much as thought of it—now accept the duties they meet; and they fall to their job of repairing an antique carryall into which three nuns, embarked upon God knows what pilgrimage, off for God knows what haven invented in a fairytale, have hustled a dozen children threatened by death.

# XV

Like Alias when he slipped his gun back into its holster, I shall not sit in judgment upon these men who threw in their hand. Where was the breath to come from that would bring them life? Where the wave that would reach them, the vision that would unite them? All that they knew of the rest of the world was contained in the crazy rumors that sprouted by the roadside every mile or two in the form of ludicrous hypotheses, and somehow, slowly spreading through a mile or two of the chaos, were transformed into certainties. The United States had declared war. The Pope had committed suicide. Russian planes had set fire to Berlin. The Armistice had been signed three days ago. Hitler had landed in England.

There is no shepherd for the women and children, but none for the men, either. The general is able to communicate with his orderly. The cabinet minister with his messenger. It may be that by their eloquence general and minister are able to transfigure their servants. Alias is able to communicate with his pilots and to win from them the sacrifice of their lives. The sergeant commanding the lorry is able to communicate with his squad. Beyond this, there is no way in the world of welding oneself to the rest. Even if we assume that at the moment of my flight towards Arras a genius existed who knew precisely what was happening in France, and that genius, that chief, had a plan that would save France, all that that chief possessed to carry out his plan was an electric cord which rang a bell in his reception room; and the army he commanded was made up of his messenger—provided that his messenger was still at his post in the reception room.

Those stray parties of soldiers who, separated from their scattered units, wandered over the jammed roads, were soldiers with no soldiering to do. But they were not filled with that despair which the vanquished patriot is supposed to feel. If in their confusion of mind they longed for peace, the peace they longed for meant to them the end of this unspeakable chaos and the return to some kind of identity, however humble. A shoemaker among them might dream that he was hammering pegs into a shoe. To hammer pegs into a shoe again would mean for him building a world. And if these men allowed themselves to be rolled back by the tide it was because the general chaos had disintegrated them, and not because they felt a horror of dying. They felt no horror of anything; they were empty of feeling.

We may take it as incontrovertible fact that men cannot be changed overnight from conquered into conquerors. Anybody who speaks of an army falling back in order to go on fighting is employing verbal subterfuge. The troops that fall back, and those that give battle, are not the same men. The army that fell back was no longer an army. I do not mean that men in retreat become unworthy of victory. Simply, the fact of falling back destroys all the ties, material and spiritual, by which they were once united. What was once an army becomes a scattering of disintegrated parts allowed to filter back to the rear. Fresh reserves are substituted for them, because the reserves constitute an organism, a whole. It is they, not a reorganized army, who undertake to block the path of the enemy. As for the fugitives, an attempt is made to collect them and re-shape them into an army. But where, as in France, there are no reserves to throw in, your initial retreat is irreparable.

There is but one principle of unity, and that is victory. Defeat not only splits men off from other men, it creates a split within the individual himself. If those apathetic fugitives do not mourn the fate of a collapsing France it is simply because they are the defeated. It is in the hearts of those men that France has been defeated. To weep for France is already the promise of victory.

Virtually none of those men, neither those still fighting nor those already benumbed, will see that whole, will see the face of a vanquished France, until later, when the tumult has died and silence has been restored. Today, in the midst of defeat, each man is concentrated upon a stubborn or shattered vulgar detail—a broken-down lorry, a road bottled up, a throttle stuck fast, a sortie that is a patent absurdity. The absurdity of the sortie is a sign of the collapse. The very act performed to arrest the collapse is a sign of absurdity. For every element stands divided against itself: there is no unity.

During a retreat there is no weeping over the universal disaster but only over the thing for which one is personally responsible. Alias, if he were to weep, would weep over the imbecility of the depots that refuse to deliver spare parts except against a requisition drawn in due form—parts which tomorrow will fall into the hands of the enemy. Collapsing France has become a deluge of fragments none of which has any identity—neither this aeroplane nor that lorry, neither that highway nor this foul throttle that refuses to budge.

Of course a collapse is a sad spectacle. Base men reveal themselves base. Pillagers reveal themselves pillagers. Institutions crumble. Troops heartsick and weary decompose. All these effects are implicit in defeat as death is implicit in the death rattle. But if the woman you loved were run over by a lorry, would you feel impelled to criticise her ugliness?

The injustice of defeat lies in the fact that its most innocent victims are made to look like heartless accomplices. It is impossible to see behind defeat the sacrifices, the austere performance of duty, the self-discipline and the vigilance that are there—those things the god of battle does not take account of. Defeat cannot show love, though love is there. Defeat shows up generals without authority, men without organization, crowds that are passive. Unquestionably, slackers and cowards have their part in this defeat. But what do they signify? What is really significant is that the rumor of a Russian change of heart or an American intervention was enough to triple the value of those men. Enough to bind them together again in a common hope. Each time that such a rumor blew through France like a sea wind, our men were filled with a fresh exaltation. If France is to be judged, judge her not by the effects of her defeat but by her readiness to sacrifice herself.

In accepting the challenge of this war, France accepted the risk of disfigurement for a time by defeat. Was France to refuse this challenge, which is to say, this risk of disfigurement? There were Frenchmen who said: "We cannot in a single year create the forty million Frenchmen needed to match those eighty millions of Germans. We cannot overnight transform a nation of farmers into a people of factory workers such as the Germans are. We cannot change our wheatfields into coalfields. We cannot look for American intervention. The Germans demand Danzig. They thus impose upon us, not the duty of saving Danzig, which is impossible, but of committing suicide in order to preserve our honor. Why? What dishonor is there in possessing a land that brings forth more wheat than machines? What dishonor is there in being only forty millions to the other man's eighty millions? Why should the dishonor be ours, and not the whole world's?" They were perfectly right. War, for France, signified disaster. Was France to refuse to fight in order to spare herself defeat? I think not. And France must instinctively have thought the same, since these warnings could not dissuade France from war. Among us, spirit conquered intelligence.

Life always bursts the boundaries of formulas. Defeat may prove to have been the only path to resurrection, despite its ugliness. I take it for granted that to create a tree I condemn a seed to rot. If the first act of resistance comes too late it is doomed to defeat. But it is, nevertheless, the awakening of resistance. Life may grow from it as from a seed.

France played her part. Her part consisted in offering herself up to be crushed and in seeing herself buried for a time in silence—since the world chose to arbitrate, and neither fought nor united against a common enemy. When a fort is to be taken by storm some men necessarily are in the front rank. Almost always, those men die. But the front rank must die if the fort is to be captured.

Since we of France agreed to fight this war without illusions, this was the rôle that fell to us. We put farmers into the field against factory workers; one man into the field against three. And who is to sit in judgment upon the ugliness of the collapse? Is a pilot brought down in flames to be judged by the consequences? Obviously, he will be disfigured.

## XVI

Which does not prevent this from being a funny war—aside from the spiritual reality that made it necessary. A funny war! I was never ashamed of this label. Hardly had we declared war when, being in no state to take the offensive, we began to look forward to our annihilation. Here it is.

We set up our haycocks against their tanks; and the haycocks turned out useless for defence. This day, as I fly to Arras, the annihilation has been consummated. There is no longer an army, there is no liaison, no matériel, and there are no reserves.

Nevertheless I carry on as solemnly as if this were war. I dive towards the German army at five hundred miles an hour. Why? I know! To frighten the Germans. To make them evacuate France. For since the intelligence I may bring back will be useless, this sortie can have no other purpose.

A funny war!

As a matter of fact, I am boasting. I have lost a great deal of altitude. Controls and throttles have thawed out. I have stepped down my speed to no more than three hundred and thirty miles an hour. A pity. I shall frighten the German army much less.

After all, it is we ourselves who call this a funny war. Why not? I should image that no one would deny us the right to call it that if we please, since it is we who are sacrificing ourselves, not those others who think our epithet immoral. Surely I have the right to joke about my death if joking about it gives me pleasure. And Dutertre has that right. I have the right to play with paradoxes. Why is it that those villages are still in flames? Why is it that that population has poured pell-mell out on the pavements? Why is it that we rush with inflexible determination towards an unmistakable slaughter-house?

I have every right to my joke; for in this moment I am fully conscious of what I am doing. I accept death. It is not danger that I accept. It is not combat that I accept. It is death. I have learnt a great truth. War is not the acceptance of danger. It is not the acceptance of combat. For the combatant, it is at certain moments the pure and simple acceptance of death.

And while men in the outside world were wondering, "Why is it that more Frenchmen are not being killed?" I was wondering, as I watched our crews go off to their death, "What are we giving ourselves to? Who is still paying this bill?"

For we were dying. For one hundred and fifty thousand Frenchmen were already dead in a single fortnight. Those dead do not exemplify an extraordinary resistance. I am not singing the praises of an extraordinary resistance. Such a resistance was impossible. But there were clusters of infantrymen still giving up their lives in undefendable farmhouses. There were aviation crews still melting like wax flung into a fire.

Look once again at Group 2-33. Will you explain to me why, as I fly to Arras, we of Group 2-33 still agree to die? For the esteem of the world? But esteem implies the existence of a judge. And I have the impression that none of us will grant whoever it may be the right to sit in judgment. To us who imagine that we are defending a cause which is fundamentally the common cause, the cause of Poland, of Holland, of Belgium, of Norway; to us who hold this view, the rôle of arbiter seems much too comfortable. It is we who sit in judgment upon the arbiter. I invite you to try to explain to us who take off with a "Very good, sir," having one chance in three to get back when the sortie is an easy one; I invite you to try to explain to a certain pilot out of another group, half of whose neck and jaw were shot away so that he is forced to renounce the love of woman for life, is frustrated in a fundamental right of man, frustrated as totally as if he were behind prison walls, surrounded inescapably by his virtue and preserved totally by his disfigurement, isolated completely by his ugliness—I invite you to explain to him that spectators are sitting in judgment upon him. Toreadors live for the bull-fight crowd: we are not toreadors. If you said to another of my friends, to Hochedé, "You've got to go up because the crowd have their eye on you," Hochedé would answer:

"There must be some mistake: it is I, Hochedé, who have my eye on the crowd."

For after all, why do we go on fighting? For democracy? If we die for democracy then we must be one of the democracies. Let the rest fight with us, if that is the case. But the most powerful of them, the only democracy that could save us, chooses to bide its time. Very good. That is its right. But by so doing, that democracy signifies that we are fighting for ourselves alone. And we go on fighting despite the assurance that we have lost the war. Why, then, do we go on dying? Out of despair? But there is no despair. You know nothing at all about defeat if you think there is room in it for despair.

There is a verity that is higher than the pronouncements of the intelligence. There is a thing which pierces and governs us and which cannot be grasped by the intelligence. A tree has no language. We are a tree. There are truths which are evident, though not to be put into words. I do not die in order to obstruct the path of the invasion, for there is no shelter upon which I can fall back with those I love. I do not die to preserve my honor, since I deny that my honor is at stake, and I challenge the jurisdiction of my judge. Nor do I die out of desperation.

And yet Dutertre, looking at his map, having pin-pointed the position of Arras somewhere round the one hundred and seventy-fifth degree of the compass, is about to say to me—I can feel it:

"175°, Captain."

And I shall accept.

# XVII

"172°."

"Right! 172°."

Call it one seventy-two. Epitaph: "Maintained his course accurately on 172°." How long will this crazy challenge go on? I am flying now at two thousand three hundred feet beneath a ceiling of heavy clouds. If I were to rise a mere hundred feet Dutertre would be blind. Thus we are forced to remain visible to the anti-aircraft batteries and play the part of an archer's target for the Germans. Two thousand feet is a forbidden altitude. Your machine serves as a mark for the whole plain. You drain the cannonade of a whole army. You are within range of every calibre. You dwell an eternity in the field of fire of each successive weapon. You are not shot at with cannon but beaten with a stick. It is as if a thousand sticks were used to bring down a single walnut.

I had given a bit of thought to this problem. There is no question of a parachute. When the stricken plane dives to the ground the opening of the escape hatch takes more seconds than the dive of the plane allows. Opening the hatch involves seven turns of a crank that sticks. Besides, at full speed the hatch warps and refuses to slide.

That's that. The medicine had to be swallowed some day. I always knew it. Meanwhile, the formula is not complicated: stick to 172°. I was wrong to grow older. Pity. I was so happy as a child. I say so; but is it true? For already in that dim hall I was moving on this same course, 172°. Because of my uncles.

Now is the time when childhood seems sweet. Not only childhood, but the whole of my past life. I see it in perspective as if it were a landscape.

And it seems to me that I myself am unalterable. I have always felt what I now feel. Doubtless my joys and sadness changed object from time to time. But the feelings were always the same. I have been happy and unhappy. I have been punished and forgiven. I have worked well and badly. That depended on the days.

What is my earliest memory? I had a Tyrolian governess whose name was Paula. But she is not even a memory: she is the memory of a memory. Paula was already no more than a legend when at the age of five I sat marooned in the dim hall. Year after year my mother would say to us round the New Year, "There is a letter from Paula." That made all the children happy. But why were we happy? None of us remembered Paula. She had gone back long before to her Tyrol. To her Tyrolian house. A house, we imagined, deep in snow and looking like the toy chalet on a Tyrolian barometer. And Paula, on sunny days, would come forth to stand in the doorway of that house like the mechanical doll over the Tyrolian barometer.

"Is Paula pretty?"

"Beautiful."

"Is it sunny in the Tyrol?"

"Always."

It was always fine weather in the Tyrol. The Tyrolian barometer sent Paula farther forward out of her doorway and on to her snow-covered lawn. Later, when I was able to write, I would be set to writing letters to Paula. They always began: "My dear Paula, I am very glad to be writing to you." The letters were a little like my prayers, for I did not know Paula.

"One seventy-four."

"Right! One seventy-four."

Call it one seventy-four. Must change that epitaph.

Strange, how of a sudden life has collected in a heap. I have packed up my memories. They will never be of use to me again. Nor to anyone else. I remember a great love. My mother would say to us: "Paula sends kisses to you all." And my mother would kiss us all for Paula.

"Does Paula know I am bigger?"

"Naturally."

Paula knew everything.

"Captain, they are beginning to fire."

Paula, they are firing at me! I glanced at the altimeter: two thousand one hundred and fifty feet. Clouds at two thousand three hundred. Well. Nothing to be done about it. What astonishes me is that beneath my cloudbank the world is not black, as I had thought it would be. It is blue. Marvellously blue. Twilight has come, and all the plain is blue. Here and there I see rain falling. Rain-blue.

"One sixty-eight."

"Right! One sixty-eight."

Call it one sixty-eight. Interesting, that the road to eternity should be zigzag. And so peaceful! The earth here looks like an orchard. A moment ago it seemed to me skeletal, inhumanly desiccated. But I am flying low in a sort of intimacy with it. There are trees, some standing isolated, others in clusters. You meet them. And green fields. And houses with red tile roofs and people out of doors. And lovely blue showers pouring all round them. The kind of weather in which Paula must have hustled us rapidly indoors.

"One seventy-five."

My epitaph has lost a good deal of its laconic dignity: "Maintained his course on 172°, 174°, 168°, 175°. . . ." I shall seem a very versatile fellow. What's that? Engine coughing? Growing cold. I shut the ventilators of the hood. Good. Time to change over to the reserve tanks. I pull the lever. Have I forgotten anything? I glance at the oil gauge. Everything shipshape.

"Beginning to get a bit nasty, Captain."

Hear that, Paula? Beginning to get nasty. And yet I cannot help being astonished by the blue of the evening. It is so extraordinary. The color is so deep. And those fruit trees, plum trees, perhaps, flowing by. I am part of the countryside now. Gone are the museum cases. I am a marauder who has jumped over the wall. I am running through the wet alfalfa, stealing plums. This is a funny war, Paula. A war nostalgic and beautifully blue. I got lost somehow, and strayed into this strange country in my old age. . . . O, no, I am not frightened. It's a little melancholy, that's all.

"Zigzag, Captain!"

Here is a new game, Paula. You kick the rudder bar with your right foot and then your left, and the anti-aircraft battery can't touch you. When I fell down I used to bruise myself and raise swellings. I am sure you used to cure me with compresses of arnica. I am going to need arnica awfully, I think. But still, you know, this evening air is marvellously blue!

Forward of my plane I saw suddenly three lance-strokes aimed at my machine. Three long brilliant vertical twigs. The paths of tracer-bullets fired from a small-calibre gun. They were golden. Suddenly in the blue of the evening I had seen the spurting glow of a three-branched candlestick.

"Captain! Firing very fast to port. Hard down!"

I kicked my rudder.

"Getting worse!"

Worse?

Getting worse; but I am seated at the heart of the world. All my memories, all my needs, all my loves are now available to me. My childhood, lost in darkness like a root, is at my disposal. My life here begins with the nostalgia of a memory. Yes, it is getting worse; but I feel none of those things I thought I should feel when facing the claws of these shooting stars.

I am in a country that moves my heart. Day is dying. On the left I see great slabs of light among the showers. They are like panes in a cathedral window. Almost within reach, I can all but handle the good things of the earth. There are those plum trees with their plums. There is that earth-smelling earth. It must be wonderful to tramp over damp earth. You know, Paula, I am going gently forward, swaying to right and left like a loaded hay wain. You think an aeroplane moves fast; and indeed it does, if you think of it. But if you forget the machine, if you simply look on, why, you are merely taking a stroll in the country.

"Arras!"

Yes. Very far ahead. But Arras is not a town. Arras thus far is no more than a red plume against a blue background of night. Against a background of storm. For unmistakably, forward on the left, an awful squall is collecting. Twilight alone would not explain this half-light. It wants blocks of clouds to filter a glow so somber.

The flame of Arras is bigger now. You wouldn't call it the flame of a conflagration. A conflagration spreads like a chancre surrounded by no more than a narrow fringe of living flesh. That red plume permanently alight is the gleam of a lamp that might be smoking a bit. It is a flame without flicker, sure to last, well fed with oil. I can feel it moulded and kneaded out of a compact substance, something almost solid that the wind stirs from time to time and bends as it bends a tree. That's it: a tree. Arras is caught up in the mesh of roots of this tree. And all the pith of Arras, all the substance of Arras, all the treasures of Arras leap, now become sap, to nourish this tree.

I can see that occasionally top-heavy flame lose its equilibrium to right or left, belch forth an even blacker cloud of smoke, and then collect itself again. But I am still unable to make out the town.

The whole war is summed up in that glow. Dutertre says that it is getting worse. Perched up forward, he can see better than I can. Nevertheless, I am astonished by a sort of indulgence shown us: this venomous plain shoots forth few stars.

Yes, but. . . .

You remember, Paula, that in the fairy-tales of our childhood there was always a knight who passed through frightful experiences before reaching the enchanted castle. He scaled glaciers, leapt across abysses, outwitted villains. And in the end the castle rose before him out of a blue plain gentler beneath the galloping hoofs than a green lawn. Already he thought himself victorious. Ah, Paula, you can't fool an old fairy-tale reader! The worst of his trials was still before him—the ogre, the dragon, the guardian of the castle.

Like that knight, I ride in the blue of the evening towards my castle of flame. And not for the first time. You had already left us when we began to play games. You missed the game called Aklin the Knight. We had invented it ourselves, for we sneered at games that other children could play. This one was played out of doors in stormy weather when, after the first flashes of lightning, we could tell from the rising smell of the earth and the sudden quivering of the leaves that the cloud was about to burst. There was a moment when the thickness in the boughs turned into a lightly soughing moss. That was the signal. Nothing could hold us back.

We would run as fast as we could from the deep end of the park towards the house, flying breathlessly across the lawns. The first drops of that rain were always scattered and heavy. The child first touched by them was beaten. Then the next. Then the third. Then the rest. He who survived longest was acknowledged the darling of the gods, the invulnerable. Until the next storm came he had the right to call himself Aklin the Knight. It was only a matter of seconds, and the result was each time a hecatomb of children.

I fly my plane, playing at being Aklin the Knight. I am running slowly and out of breath towards my castle of flame.

"Captain! Captain! I've never seen anything like it!"

Nor have I.

Where now is my vulnerability? Unknown to myself, I had been hoping. . . .

## XVIII

Despite my lack of altitude, I had been hoping. Despite the tank parks, despite the flame over Arras. Desperately, I had been hoping. I had escaped into a memory of early childhood in order to recapture the sense of sovereign protection. For man there is no protection. Once you are a man you are left to yourself. But who can avail against a little boy whose hand is firmly clasped in the hand of an all-powerful Paula? Paula, I have used thy shade as a shield.

I have used every trick in my bag. When Dutertre said to me, "It's getting worse," I used even that threat as a source of hope. We were at war: necessarily, then, there had to be evidence of war. The evidence was no more than a few streaks of light. "Is this your terrible danger of death over Arras? Don't make me laugh!"

The man condemned had imagined that the executioner would look like a pallid robot. Arrives a quite ordinary decent-appearing fellow who is able to sneeze, even to smile. The man condemned clings to that smile as to a promise of reprieve. The promise is a wraith. The headsman sneezes—and the head falls nevertheless. But who can reject hope?

I myself could not but be deceived by the smile I saw—since this whole world was snug and verdant, since the wet slate and tile shone so cordially, since from minute to minute nothing changed nor promised to change. Since we three, Dutertre, the gunner, and I, were men walking across fields, sauntering idly home without so much as the need to raise our collars, so little was it raining. Since here at the heart of the German zone nothing stood forth that was really worth telling about, whence it must follow that farther on the war need not of necessity be different to this. Since it seemed that the enemy had scattered and melted into the wide and rural plain, standing perhaps at the rate of one soldier to a house, one soldier to a tree, one of whom, remembering now and then the war, would fire. The order had been drummed into the fellows ears: "Fire on all enemy planes." But he had been day-dreaming, and the order had been dimmed by the dream. He let fly his three rounds without much expectation of results. Thus at dusk I used to shoot ducks that meant very little to me if the evening invited my soul. I would fire while talking about something else. It hardly disturbed the ducks.

It is so easy to spin fine tales to oneself. The enemy takes aim, but without firm purpose; and he misses me. Others in turn let us pass. Those who might trip us up are perhaps at this moment inhaling with pleasure the smell of the night, or lighting cigarettes, or finishing a funny story—and they let us pass. Still others, in the village where they are billeted, are perhaps dipping their tin cups into the soup. A roar rises and dies away. Friend or enemy? There isn't time for them to find out: their eyes are on the cup now filling—they let us pass. And I, whistling a tune and my hands in my pockets, do my best to walk as casually as I can through this garden forbidden to trespassers where every guard, counting on the next guard, lets us pass.

How vulnerable I was! Yet it seemed to me that my very vulnerability was a trap, a means of cajoling them: "Why fire? Your friends are sure to bring me down a little farther on." And they would shrug their shoulders: "Go break your neck somewhere else." They were leaving the chore to the next battery—because they were anxious not to miss their turn at the soup, were finishing their funny story, or were simply enjoying the evening breeze. I was taking advantage of their negligence, and I was saved by the seeming coincidence that all of them at once appeared to be weary of war. And why not? Already I was thinking vaguely that from soldier to soldier, squad to squad, village to village, I should get through this sortie. After all, what were we but a passing plane in the evening sky? Not enough to make a man raise his eyes.

Of course I hoped to get back. But I could feel at the same time that something was in the air. You are sentenced: a penalty hangs over you; but the gaol in which you are locked up continues silent. You cling to that silence. Every second that drops is like the one that went before. There is no reason why the second about to drop should change the world. Such a task is too heavy for a single second. Each second that follows safeguards your silence. Already this silence seems perpetual.

But the step of him who must come sounds in the corridor.

Something in this countryside suddenly exploded. So a log that seemed burnt out crackles suddenly and shoots forth its sparks.

How did it happen that the whole plain started up at the same moment? When spring comes, all the trees at once drop their seed. Why this sudden springtime of arms? Why this luminous flood rising towards us and, of a sudden, universal?

My first feeling was that I had been careless. I had ruined everything. A wink, a single gesture is enough to topple you from the tight-rope. A mountain climber coughs, and he releases an avalanche. Once he has released the avalanche, all is over.

We had been swaying heavily through this blue swamp already drowned in night. We had stirred up this silent slime; and now, in tens of thousands, it was sending towards us its golden bubbles. A nation of jugglers had burst into dance. A nation of jugglers was dribbling its projectiles in tens of thousands in our direction. Because they came straight at us, at first they appeared to be motionless. Like colored balls which jugglers seem not so much to fling into the air as to release upwards, they rose in a lingering ascension. I could see those tears of light flowing towards me through a silence as of oil. That silence in which jugglers perform.

Each burst of a machine gun or a rapid-fire cannon shot forth hundreds of these phosphorescent bullets that followed one another like the beads of a rosary. A thousand elastic rosaries strung themselves out towards the plain, drew themselves out to the breaking point, and burst at our height. When, missing us, the string went off at a tangent, its speed was dizzying. The bullets were transformed into lightning. And I flew drowned in a crop of trajectories as golden as stalks of wheat. I flew at the center of a thicket of lance strokes. I flew threatened by a vast and dizzying flutter of knitting needles. All the plain was now bound to me, woven and wound round me, a coruscating web of golden wire.

I leant towards the earth and saw those storied levels of luminous bubbles rising with the tardy movement of veils of fog. I saw as I stared the slow vortex of seed, swirling like the husks of threshed grain. And when I raised my head I saw on the horizon those stacks of lances. Guns firing? Not at all! I am attacked by cold steel. These are swords of light. I feel . . . certainly not in danger! Dazzled I am by the luxury that envelopes me.

What's that!

I was jolted nearly a foot out of my seat. The plane has been rammed hard; I thought. It has burst, been ground to bits. . . . But it hasn't; it hasn't. . . . I can still feel it responsive to the controls. This was but the first blow of a deluge of blows. Yet there was no sign of explosion below. The smoke of the heavy guns had probably blended into the dark ground.

I raised my head and started. What I saw was without appeal.

## XIX

I had been looking on at a carnival of light. The ceiling had risen little by little and I had been unaware of an intervening space between the clouds and me. I had been zigzagging along a line of flight dotted by ground batteries. Their tracer bullets had been spraying the air with wheat-colored shafts of light. I had forgotten that at the top of their flight the shells of those batteries must burst. And now, raising my head, I saw around and before me those rivets of smoke and steel driven into the sky in the pattern of towering pyramids.

I was quite aware that those rivets were no sooner driven than all danger went out of them, that each of those puffs possessed the power of life and death only for a fraction of a second. But so sudden and simultaneous was their appearance that the image flashed into my mind of conspirators intent upon my death. Abruptly their purpose was revealed to me, and I felt on the nape of my neck the weight of an inescapable reprobation.

Muffled as those explosions reached me, their sound covered by the roar of my engines, I had the illusion of an extraordinary silence. Those vast packets of smoke and steel moving soundlessly upward and behind me with the lingering flow of icebergs, persuaded me that, seen in their perspective, I must be virtually motionless. I was motionless in the dock before an immense assizes. The judges were deliberating my fate, and there was nothing I could plead. Once again the timelessness of suspense seized me. I thought,—I was still able to think,—"They are aiming too high," and I looked up in time to see straight overhead, swinging away from me as if with reluctance, a swarm of black flakes that glided like eagles. Those eagles had given me up. I was not to be their prey. But even so, what hope was there for me?

The batteries that continued to miss me continued also to readjust their aim. New walls of smoke and steel continued to be built up round me as I flew. The ground-fire was not seeking me out, it was closing me in.

"Dutertre! How much more of this is there?"

"Stick it out three minutes, Captain. Looks bad, though."

"Think we'll get through?"

"Not on your life!"

There never was such muck as this murky smoke, this mess as grimy as a heap of filthy rags. The plain was blue. Immensely blue. Deep-sea blue.

What was a man's life worth between this blue plain and this foul sky? Ten seconds, perhaps; or twenty. The shock of the exploding shells set all the sky shuddering. When a shell burst very near, the explosion rumbled along the plane like a rock dropping through a chute. And when for a moment the roar stopped, the plane rang with a sound that was almost musical. Like a sigh, almost; and the sigh told us that the plane had been missed. Those bursts were like the thunder: the closer they came, the simpler they were. A rumble meant distance, a clean *bang!* meant that we had been squarely hit by a shell fragment. The tiger does not do a messy job on the ox it brings down. The tiger sets its claws into the ox without skidding. It takes possession of the ox. Each square hit by a fragment of shell sank into the hull of the plane like a claw into living flesh.

"Anybody hurt?"

"Not I!"

"Gunner! You all right?"

"O.K., sir!"

Somehow those explosions, though I find I must mention them, did not really count. They drummed upon the hull of the plane as upon a drum. They pierced my fuel tanks. They might as easily have drummed upon our bellies, pierced them instead. What is the belly but a kind of drum? But who cares what happens to his body? Extraordinary, how little the body matters.

There are things that we might learn about our bodies in the course of everyday living if we were not blind to patent evidence. It takes this rain of upsurging streamers of light, this assault by an army of lances, this assizes set up for the last judgment, to teach us those things.

I used to wonder as I was dressing for a sortie what a man's last moments were like. And each time, life would give the lie to the ghosts I evoked. Here I was, now, naked and running the gauntlet, unable so much as to guard my head by arm or shoulder from the crazy blows raining down upon me. I had always assumed that the ordeal, when it came, would be an ordeal that concerned my flesh. My flesh alone, I assumed, would be subjected to the ordeal. It was unavoidable that in thinking of these things I should adopt the point of view of my body. Like all men, I had given it a good deal of time. I had dressed it, bathed it, fed it, quenched its thirst. I had identified myself with this domesticated animal. I had taken it to the tailor, the surgeon, the barber. I had been unhappy with it, cried out in pain with it, loved with it. I had said of it, "This is me." And now of a sudden my illusion vanished. What was my body to me? A kind of flunkey in my service. Let but my anger wax hot, my love grow exalted, my hatred collect in me, and that boasted solidarity between me and my body was gone.

Your son is in a burning house. Nobody can hold you back. You may burn up; but do you think of that? You are ready to bequeath the rags of your body to any man who will take them. You discover that what you set so much store by is trash. You would sell your hand, if need be, to give a hand to a friend. It is in your act that you exist, not in your body. Your act is yourself, and there is no other you. Your body belongs to you: it is not you. Are you about to strike an enemy? No threat of bodily harm can hold you back. You? It is the death of your enemy that is you. You? It is the rescue of your child that is you. In that moment you exchange yourself against something else; and you have no feeling that you lost by the exchange. Your members? Tools. A tool snaps in your hand: how important is that tool? You exchange yourself against the death of your enemy, the rescue of your child, the recovery of your patient, the perfection of your theorem. Here is a pilot of my Group wounded and dying. A true citation in general orders would read: "Called out to his observer, 'They've got me! Beat it! And for God's sake don't lose those notes!' " What matters is the notes, the child, the patient, the theorem. Your true significance becomes dazzlingly evident. Your true name is duty, hatred, love, child, theorem. There is no other you than this.

The flames of the house, of the diving plane, strip away the flesh; but they strip away the worship of the flesh too. Man ceases to be concerned with himself: he recognizes of a sudden what he forms part of. If he should die, he would not be cutting himself off from his kind, but making himself one with them. He would not be losing himself, but finding himself. This that I affirm is not the wishful thinking of a moralist. It is an everyday fact. It is a commonplace truth. But a fact and a truth hidden under the veneer of our everyday illusion. Dressing and fretting over the fate that might befall my body, it was impossible for me to see that I was fretting over something absurd. But in the instant when you are giving up your body, you learn to your amazement—all men always learn to their amazement—how little store you set by your body. It would be foolish to deny that during all those years of my life when nothing insistent was prompting me, when the meaning of my existence was not at stake, it was impossible for me to conceive that anything might be half so important as my body. But here in this plane I say to my body (in effect), "I don't care a button what becomes of you. I have been expelled out of you. There is no hope of your surviving this, and yet I lack for nothing. I reject all that I have been up to this very instant. For in the past it was not I who thought, not I who felt: it was you, my body. One way and another, I have dragged you through life to this point; and here I discover that you are of no importance."

Already at the age of fifteen I might have learnt this lesson. I had a younger brother who lay dying. One morning towards four o'clock his nurse woke me and said that he was asking for me.

"Is he in pain?" I asked.

The nurse said nothing, and I dressed as fast as I could.

When I came into his room he said to me in a matter-of-fact

voice, "I wanted to see you before I died. I am going to die." And with that he stiffened and winced and could not go on. Lying in pain, he waved his hand as if saying "No!" I did not understand. I thought it was death he was rejecting. The pain passed, and he spoke again. "Don't worry," he said. "I'm all right. I can't help it. It's my body." His body was already foreign territory, something not himself.

He was very serious, this younger brother who was to die in twenty minutes. He had called me in because he felt a pressing need to hand on part of himself to me. "I want to make my will," he said; and he blushed with pride and embarrassment to be talking like a grown man. Had he been a builder of towers he would have bequeathed to me the finishing of his tower. Had he been a father, I should have inherited the education of his children. A reconnaissance pilot, he would have passed on to me the intelligence he had gleaned. But he was a child, and what he confided to my care was a toy steam engine, a bicycle, and a rifle.

Man does not die. Man imagines that it is death he fears; but what he fears is the unforeseen, the explosion. What man fears is himself, not death. There is no death when you meet death. When the body sinks into death, the essence of man is revealed. Man is a knot, a web, a mesh into which relationships are tied. Only those relationships matter. The body is an old crock that nobody will miss. I have never known a man to think of himself when dying. Never.

"Captain!"
"What's up?"
"Getting hot!"
"Gunner!"
"Er . . . yes, sir."

"What—."

My question vanished in the shock of another explosion.

"Dutertre!"

"Captain?"

"Hurt?"

"No."

"You, gunner!"

"Yes, sir."

"I wa—."

I seemed to be running the plane into a bronze wall. A voice in my ear said, "Boy! oh, boy!" as I looked up to measure the distance to the overhanging clouds. The sharper the angle at which I stared, the more densely the murky tufts seemed to be piled up. Seen straight overhead, the sky was visible between them, and they hung curved and scattered, forming a gigantic coronet in the air.

A man's thigh muscles are incredibly powerful. I bore down upon the rudder with all my strength and sent the plane shuddering and skidding at right angles to our line of flight. The coronet swung overhead and slid down on my right. I had got away from one of the batteries and left it firing wasted packets of shell. But before I could bring my other thigh into play the ground battery had set straight what hung askew—the coronet of smoke was back again. Once more I bore down, and again the plane groaned and swayed in this swampy sky. All the weight of my body was on that bar, and the machine had swung, had skidded squarely to starboard. The coronet curved now above me on the left.

Would we last it out? But how could we! Each time that I brought the ship brutally round, the deluge of lance-strokes followed me before I could jerk back again. Each time the coronet was set back into place and the shell bursts shook up the plane anew. And each time, when I looked down, I saw again that same dizzyingly slow ascension of golden bubbles that seemed to be accurately centered upon my plane. How did it happen that we were still whole? I began to believe in us. "I am invulnerable, after all," I said to myself, "I am winning. From second to second, I am more and more the winner."

"Anybody hurt yet?"

"Nobody."

They were unhurt. They were invulnerable. They were victorious. I was the owner of a winning team. And from that moment each explosion seemed to me not to threaten us but to temper us. Each time, for a fraction of a second, it seemed to me that my plane had been blown to bits; but each time it responded anew to the controls and I nursed it along like a coachman pulling hard on the reins. I began to relax, and a wave of jubilation went through me. There was just time enough for me to feel fear as no more than a physical stiffening induced by a loud crash, when instantly after each buffet a wave of relief went through me. I ought to have felt successively the shock, then the fear, then the relief; but there wasn't time. What I felt was the shock, then instantly the relief. Shock, relief. Fear, the intermediate step, was missing. And during the second that followed the shock I did not live in the expectancy of death in the second to come, but in the conviction of resurrection born of the second just passed. I lived in a sort of slipstream of joy, in the wake of my jubilation. A prodigiously unlooked-for pleasure was flowing through me. It was as if, with each second that passed, life was being granted me anew. As if with each second that passed my life became a thing more vivid to me. I was living. I was alive. I was still alive. I was the source of life itself. I was thrilled through with the intoxication of living. "The heat of battle" is a familiar phrase; the heat of living is a truer one. "I wonder," I said to myself, "if those Germans below who are firing at us know that they are creating life within us?"

All my tanks had been pierced, both gas and oil. Otherwise we seemed to be sound. Dutertre called out that he was through, and once again I looked up and calculated the distance to the clouds. I raised the nose of the ship, and once again I sent the plane zigzagging as I climbed. Once again I cast a glance earthwards. What I saw I shall not forget. The plain was crackling everywhere with short wicks of spurting flame—the rapid-fire cannon. The colored balls were still floating upward through an immense blue aquarium of air. Arras was glowing dark red like iron on the anvil, a flame fed by subterranean stores, by the sweat of men, the inventions of men, the arts of men, the memories and patrimony of men, all these braided in the ruddy ascension of that single plume that changed them into fire and ash, borne away on the wind.

Already I was flying through the first packets of mist. Golden arrows still rose and pierced the belly of the cloud, and just as the cloud closed round me I caught through an opening my last glimpse of that scene. For a single instant the flame over Arras rose up glowing in the night like a lamp in the nave of a cathedral. The lamp that was Arras was burning in the service of a cult, but at a price. By to-morrow it would have consumed Arras and itself have been consumed.

"Everything all right, Dutertre?"

"First-rate, Captain. Two-forty, please. We shan't be able to come down out of this cloud for about twenty minutes. Then I'll pick up a landmark along the Seine somewhere."

"Everything all right, gunner?"

"Everything fine, sir."

"Not too hot for you, was it?"

"No, I guess not, sir."

Hard for him to tell. But he was feeling fine. I thought of Gavoille's gunner. In the days when this was still a very odd war, we used to do long-distance reconnaissance over Germany. There was a night over the Rhine when eighty searchlights picked up Gavoille's plane and built a giant basilica round it. The anti-aircraft began to fire, and suddenly Gavoille heard his gunner talking to himself—for the inter-com is hardly a private line. The man was muttering a dialogue of one: "Think you've been around, do you? I'll tell you something you've never seen!" He was feeling fine, that gunner.

I flew on, drawing deep slow breaths. I filled my lungs to the bottom. It was wonderful to breathe again. There were many things I was going to find out about. First I thought of Alias. No, that's not true. I thought first of my host, my farmer. I still looked forward to asking him how many instruments he thought a pilot had to watch. Sorry, but I am stubborn about some things. One hundred and three. He would never guess. Which reminds me. When your tanks have been pierced, it does no harm to have a look at your gauges. Wonderful tanks! Their rubber coatings had done their job; automatically, they had contracted and plugged the holes made by bullets and shell splinters. I had a look at my stabilizers. This cloud we flew in was a storm cloud. It shook us up pretty badly.

"Think we can come down now?"

"Ten minutes more. Better wait another ten minutes."

Of course I could wait another ten minutes. . . . Yes, I had thought of Alias. Was he still expecting us, I wondered? The other day we had been half an hour late. A half hour is generally longer than you ought to be: it means trouble. I had landed and run to join the Group, who were at table. I had opened the door and fallen into a chair beside Alias. At that moment he had a cluster of spaghetti on his fork and was preparing to tuck it away. He jumped, took a good look at me, and sat perfectly still, the noodles hanging from his fork.

"Well, I . . . Glad to see you," he said.

And he stuffed the noodles into his mouth.

The major has one serious fault, to my mind. He insists stubbornly on examining his pilots about their sorties. He will examine me. He will sit looking at me with embarrassing patience, waiting for me to spin out my commonplace observations. He will have armed himself with paper and pencil, determined not to lose a single drop of the elixir I shall presumably have brought back.

I thought of school: "Saint-Exupéry, how do you integrate Bernoulli's equations?"

"Er . . . er."

Bernoulli, Bernoulli. Let me see. . . . And you stiffened under the teacher's gaze, motionless, fixed in place like an insect on a pin.

Intelligence is Dutertre's business, not mine. He is the observer; I am the pilot. From where he sits he can see straight below. He sees lots of things—lorries, barges, tanks, soldiers, cannon, horses, railway stations, trains, station masters. From where I sit I see the world at an angle. I see clouds, sea, rivers, mountains, sun. I see roughly, and get only a general impression.

"Major, you know as well as I do that a pilot . . ."

"Come, come, Saint-Ex! You do see some things, after all."

I . . . Oh, yes! Flames. Villages burning. Doesn't the major think that interesting?

"Nonsense! The whole country is on fire. What else?"

Why must Alias be so cruel?

# XX

What I bring back from this sortie is not matter for a report. When Alias examines me I shall flunk like a schoolboy standing before all the class at a blackboard. I shall seem very unhappy, and yet I shall not be unhappy. Unhappiness is behind me. It fled in that instant when the shell bursts began to drum upon the plane. Had I turned back one second before, I should have missed knowing myself.

I should never have known the flood of affection that at this moment fills my heart. I am going back to my own kind. I am going home. I am like a housewife whose shopping is done and who is on her way home, her mind on the savoury dinner with which she is about to delight her family. Her market basket swings on her arm to left and right. From time to time she raises the newspaper that covers it, and peers in. Everything is there: nothing has been forgotten. She smiles to herself at the thought of the surprise she is planning. She lingers a little, glances into the shop windows.

I too should be glancing into my shop windows if Dutertre did not insist that I go on in this pallid prison of cloud. I should be glancing at the passing countryside. Though Dutertre is right to insist that I be patient. This area over which I fly is treacherous: its air is heavy with conspiracy. Each little manor house below, with its slightly ridiculous lawn and handful of domesticated trees standing like an artless background for a family photograph, has become a blind. If I were to fly low over those houses, no friendly hands would wave to me, but shells would rise and explode.

Yet even in the belly of this cloud I am on my way home from market. The major was right, after all. When he sent us off in a voice that seemed to say, "And then you take the second turn to the right, where you will see a tobacco shop," his voice was pitched on the right note. My conscience is at rest. I have the major's matches in my pocket—or more truly, Dutertre has them in his pocket. How Dutertre manages to remember what he saw, I cannot imagine. But that is his business. My mind is on more serious things. We shall land; and if the enemy spare us the nuisance of a sudden rush to still another field, I shall challenge Lacordaire and beat him at chess. He hates to lose. So do I. But I shall win.

Yesterday, be it said without dishonor, Lacordaire got tight. At least, a little tight. He had got tight in order to console himself. Coming in from a sortie, he had forgotten to release his landing gear and had set the plane down on her belly. Unfortunately, Alias had seen him do it; but he had not said a word. And Lacordaire, a pilot of long experience, had stood by, waiting for Alias to turn upon him. He had stood by hoping that Alias would curse him. A violent tongue lashing would have done him good. It would have allowed him to explode too. It would have allowed him to get off his chest the rage against himself that was swelling in him. But Alias had merely shaken his head sadly. Alias' mind was on the plane, not on the pilot. To the major, this accident was a kind of anonymous misfortune, a statistical tax levied on the Group. It was the effect of one of those moments of absent-mindedness that attack even the most experienced pilots. It was an injustice, and Lacordaire was its victim. Except this blunder, Lacordaire's professional record was clean. Alias knew this, and all that bothered him was the plane. Automatically, without thinking, he turned to Lacordaire and asked him how bad he thought the damage was. And I could feel Lacordaire's pent up rage rise a degree at the question. You put your hand cordially on the torturer's shoulder and say to him, "How badly do you think your victim is suffering?" Truly, the human heart is unfathomable. That friendly hand soliciting the torturer's sympathy exasperates the torturer. He flings a black look at the victim and is sorry he hasn't finished her off.

I am on my way home. Group 2-33 is my home. And I understand the men who live in my home. I cannot be mistaken about Lacordaire. Lacordaire cannot be mistaken about me. Nothing is stronger than the community of feeling between us, the feeling that goes through me when I say, "We of Group 2-33." The particles, the fragments that we are, collect and possess meaning in the fact of the Group.

Flying in the cloud, I think of Gavoille and Hochedé. I am stirred by the community of feeling that binds me to them. I wonder about Gavoille. What sort of people does he come of? There is a wonderful earthy substance in Gavoille. A memory sweeps suddenly over me and fills me with warmth. At Orconte, Gavoille too was billeted with a peasant. One day he said to me, "The farmer's wife killed a pig the other day. She wants us to try her blood-sausage."

Three of us sat eating the wonderful black and crackling skin—Gavoille, Israel, and I. There was a crock of white wine to wash it down. Gavoille said as we ate, "I bought this for the farmer's wife, thinking she'd like it. Write something in it for her." It was a copy of one of my books. I was not in the least embarrassed. I wrote in it with pleasure, to please them both. Gavoille sat scratching his leg. Israel was stuffing his pipe. The farmer's wife seemed pleased to have a book inscribed by an author. The kitchen was redolent of the sausage. I was a little tight, for the white wine was heady. I did not feel in the least strange, despite the fact of inscribing a book—a thing which in other circumstances has always bothered me. I did not feel at all out of place. Despite the book, I did not think of myself either as an author or as an outsider. I was not an outsider. Israel looked on and smiled pleasantly as I wrote my name. Gavoille went on scratching his leg. And I felt grateful for the way they took it. That book might have made them look upon me as an outsider. Yet it didn't. I was still one of them.

The notion of looking on at life has always been hateful to me. What am I if I am not a participant? In order to be, I must participate. I am fed by the quality that resides in those who participate with me. That quality is something the men of the Group never think of—not out of humility, but because they do not stoop to measure it. Gavoille does not wonder about himself, nor does Israel. Each of these men is a web woven of his job, his trade, his duty. That smoking sausage, eaten in these circumstances, is woven into that web. The presence of these men is dense, full of meaning, and it warms my heart. I am able to sit with them in silence. To drink my white wine with them. To sign my book without thereby cutting myself off from them. Nothing in the world is strong enough to wreck this fellowship.

I do not mean to belittle the workings of the mind or the products of the intelligence. I admire a limpid intelligence as much as any man. But what is a man if he lacks substance? If he is a mere intellect and not a being? As formerly I saw substance in Guillaumet, so now I see it in Gavoille, in Israel.

I have mentioned before that because I was a writer I might have enjoyed certain advantages, certain liberties in this war. I might for example have been free to leave Group 2-33 the day I no longer approved of what I was ordered to do. But that kind of liberty I reject almost with terror. It is no more than the liberty to be a bystander, which is to say the liberty not to exist. There is no growth except in the fulfillment of obligations.

We in France all but died of intelligence unsupported by substance. Gavoille exists. He loves, hates, rejoices, complains. He is shaped and heightened by the strands woven together and constituting his being. And exactly as, sitting with him at table, I took pleasure from the crisp sausage we shared, so I take pleasure from the obligations of the craft that fuse us of the Group into a common being. I love Group 2-33. I do not love it with the love of a spectator looking on at a handsome spectacle. I don't give a button for spectacles. I love Group 2-33 because I am part of it and it is part of me, because it nourishes me and I contribute to nourishing it.

And now, flying home from Arras, I am more than ever interwoven with Group 2-33. I have formed still another tie with it. I have intensified in me that feeling of communion with it that is to be relished and left unspoken. Each of us had risked his life in more or less the same fashion. Israel had disappeared. It seemed pretty certain that in the course of to-day's outing I too should disappear. What have I earned by this swing round the sky except a slightly better right to sit down at their table and be silent with them? The right is dearly bought; but it is a dear right. It is the right to be, and thus to escape non-being.

Yet the notion that I shall stammer when, some minutes from now, Alias will put his questions, makes me go red. I shall feel ashamed, I know. The major will think me a little idiotic. The shame that I feel already by anticipation is genuine. Yet . . . Once again I had taken off—this time to Arras—in search of the proof of my good faith. I had risked my flesh in this sortie. I had risked it being pretty sure that I should lose it. I had given everything to the rules of the game in order to turn them somehow into something other than the rules of the game. And this being so, I have won the right to appear sheepish when the major examines me. The right, that is, to participate. To be interwoven with the rest. To commune with them. To give and receive. To be more than myself. To possess this plenitude that swells so powerfully within me. To feel the love that I feel for the Group, a love that is not an impulse from without but is something inward and never to be manifested—except at a farewell dinner. At a farewell dinner you are sure to be a little drunk, and the benevolence born of alcohol is sure to make you lean towards your friends as a tree whose boughs bend with gifts. My love of the Group has no need of definition. It is woven of bonds. It is my substance. I am of the Group, and the Group is of me.

And as I think of the Group, it is impossible for me not to think of Hochedé. Hochedé made a total gift of himself to this war. More, probably, than any of us, Hochedé dwells permanently in that state which I have striven so hard to attain to. Hochedé has arrived at the goal towards which the rest of us tend, the goal I seek to reach.

Hochedé is a former sergeant recently promoted second lieutenant. I can imagine that his culture is slight. He is unable to shed any light upon himself. But he is constructed, he is complete. The word duty loses all bombast when applied to Hochedé. Any man would be happy to accept his duty as Hochedé does. When I think of Hochedé, I reproach myself all my petty renunciations, my negligences, my laziness, and my moments of intellectualism, which is to say scepticism. This is not a sign in me of virtue but of intelligent jealousy. I should like to exist as completely as Hochedé does. A tree solidly planted on its roots is a beautiful thing. The permanence of Hochedé is a beautiful thing. Hochedé could never disappoint.

Volunteer? We were all volunteers on all our sorties. For the rest of us, the reason was a vague need to believe in ourselves. By volunteering, we outdid ourselves a little. Hochedé was a volunteer by nature. He was, in essence, this war. The fact was so evident that when a plane was bound to be sacrificed the major thought automatically of Hochedé. "Look here, Hochedé. . . ." Hochedé was steeped in this war as a monk is steeped in religion. For whom did he fight? For himself, since he was interwoven with the war, with the Group, with France. Hochedé was fused together with a certain substance, and that substance, which was his own significance, had to be saved. At Hochedé's level, life and death are somewhat the same thing. Hochedé was already part of both. Perhaps without realizing it, he hardly feared death. Stick it out; make others stick it out—that was what mattered. For Hochedé, life and death had become reconciled.

The first time that Hochedé amazed me was when, he being still a sergeant, Gavoille tried to borrow his stop-watch in order to clock a ship.

"Lieutenant, sir. I. . . . I'd rather not lend it."

"Don't be a fool! I'll give it back to you in ten minutes."

"Sir, there's a stop-watch at the squadron depot."

"Yes, a broken one. I know it."

"Sir, I . . . Nobody lends stop-watches, sir. I don't have to lend it to the lieutenant. It's not in orders. The lieutenant hasn't the right to insist."

Military discipline and respect for the hierarchy may demand that a Hochedé just unlimbered from a parachute and out of a burning plane jump instantly into another plane and take off on a sortie twice as dangerous. It may not demand that he turn over to an officer a stop-watch that has cost him three months' pay and that seems to him as precious and fragile as a baby. You can tell by the way some men wave their arms that they have no respect for stop-watches. Gavoille seemed to Hochedé just such a man. And when Hochedé, still fuming with indignation, but having won out, left the room with his stop-watch over his heart, I could have embraced him. Hochedé was a man with a heart. He would fight to the death for his stop-watch. His stop-watch existed. He would die for his country. His country existed. Hochedé existed, being interwoven with both. He was shaped and heightened by his ties with watch and country.

And so Hochedé was precious in my eyes, though there was no need to tell him so. For like reasons, when Guillaumet, the best friend I ever had, was killed in the course of duty, there was no need for me to speak of him. We had flown the same airlines. Participated in the building of the same structures. Were of the same substance. Something of me died in him. Guillaumet became one of the companions of my silence. I am part of Guillaumet, and Guillaumet is part of me.

I am part of Guillaumet, of Gavoille, of Hochedé, and they are part of me. I am part of Group 2-33, and it of me. I am part of my country, and it of me. My country and I are one. And all the men of Group 2-33 are one with their country.

## XXI

I have changed a good deal. I had been bitter these last days, Major Alias—these last days when the armored invasion was meeting no resistance, when our sacrificial offerings cost the Group seventeen out of twenty-three crews. It had seemed to me that we —that you in particular—were agreeing to play the part of dead men merely because the show called for dead supernumeraries. I had been bitter, Major Alias; and I had been wrong.

You in particular, but the rest of us too, had clung to the letter of a duty whose spirit had ceased to be visible for us. You had driven us intuitively not towards victory, which was impossible, but towards self-fulfillment. You knew as well as we did that the intelligence we brought back would never reach the Staff. But you were salvaging rites whose power none of us could perceive. Each time that you examined us on the lorries, the barges, the railway trains we had spotted, examined us as soberly as if our answers could possibly serve a purpose, you seemed to me revoltingly hypocritical. But you were right, Major Alias.

Until I learnt what I learnt over Arras, I could feel no responsibility for this stream of refugees over which once more I fly. I can be bound to no men except those to whom I give. I understand no men except those to whom I am bound. I exist only to the degree that I am nourished by the springs at my roots. I am bound to that mob on the highways, and it is bound to me. At three hundred miles an hour and an elevation of six hundred feet, now that I have come down out of the clouds, I have become one with that mob. I, flying in the descending night, am like a shepherd who in a single glance counts and collects and welds his scattered sheep into a flock again. That mob is no longer a mob, it is a people.

We dwell in the rot of defeat, yet I am filled with a solemn and abiding jubilation, as if I had just come from a sacrament. I am steeped in chaos, yet I have won a victory. Is there a single pilot of the Group who ever flew home without this feeling of victory in his breast? This very day, when Pénicot came in from a morning's low-altitude sortie and was telling me about it, this was how he spoke: "Whenever one of their ground batteries seemed to me to be aiming too well for my comfort, I would zoom down just above the ground and make straight for the battery at full speed, and the spray from my guns would blow out their ruddy fire as if it was a candle. Before they knew it, I was on their gun crew, and you would have thought I was a bursting shell. *Bang!* The crew would scatter and flop in every direction. I swear, I felt as if I was scattering nine-pins." And Pénicot, victorious captain, roared with glee, as pleased with himself as Gavoille's gunner when they flew through the vault of the enemy searchlights like a military wedding-party marching under an arch of swords.

"Ninety-four, captain."

Dutertre had picked up a landmark along the Seine, and we were down now to four hundred feet. Flowing beneath me at three hundred miles an hour, the earth was drawing great rectangles of wheat and alfalfa, great triangles of forest, across my glass windscreen. Divided by the stem of the plane, the flow of the broken landscape to left and right filled me with a curious satisfaction. The Seine shone below, and when I crossed its winding course at an angle it seemed to speed past and pivot upon itself. The swirl of the river was as lovely in my sight as the curve of a sickle in a field. I felt restored to my element. I was captain of my ship. The fuel tanks were holding out. I should certainly win a drink at poker dice from Pénicot and then beat Lacordaire at chess. That was how I was when my team had won.

"Captain! Firing at us! We are in forbidden territory."

Forbidden, that is, by our own people. A rectangle in which our own people fired on any plane, friend or enemy. We had orders to fly round it, but the Group never bothered to observe these traffic regulations. Well, it was Dutertre who set the course, not I. Nobody could blame me.

"Firing hard?"

"Doing as well as they can."

"Want to go back and round?"

"Oh, no."

His tone was matter-of-fact. We had been through our storm. For men like us, this anti-aircraft fire was a mere April shower. Still. . . .

"Dutertre, wouldn't it be silly to be brought down by our own guns?"

"They won't bring anything down. Just giving themselves a little exercise."

Dutertre was in a sarcastic mood. Not I. I was happy. I was impatient to be back with the Group again.

"They are, for a fact. Firing like. . . ."

The gunner! Come to, has he? This is the first time on board that he has opened his mouth without being spoken to. He took in the whole jaunt without feeling the need of speech. Unless that was he who muttered "Boy! oh, boy!" when the shells were thickest. But you wouldn't call that blabbing, exactly. He spoke now because machine guns are his specialty—and how can you keep a specialist quiet about his specialty?

It was impossible for me not to contrast in my mind the two worlds of plane and earth. I had led Dutertre and my gunner this day beyond the bourne at which reasonable men would stop. We had seen France in flames. We had seen the sun shining on the sea. We had grown old in the upper altitudes. We had bent our glance upon a distant earth as over the cases of a museum. We had sported in the sunlight with the dust of enemy fighter planes. Thereafter we had dropped earthward again and flung ourselves into the holocaust. What we could offer up, we had sacrificed. And in that sacrifice we had learnt even more about ourselves than we should have done after ten years in a monastery. We had come forth again after ten years in a monastery.

And in the little time we had taken to wander so far, the caravan of refugees over which we flew had perhaps advanced five hundred yards. In less time than it would take them to lift a motor-car out of a ditch and set it back on the road again, in less time than many a driver would sit drumming impatiently on the wheel as he waited for a stream of traffic to empty itself out of a crossroad, we should be safely back in our haven.

At a single bound we had leapt over the whole defeat. We were above and beyond it, pilgrims stronger than the desert through which they toil because already in their hearts they have reached the holy city that is their destination. This night now falling would park that unhappy people of refugees in its stable of misery. The flock would huddle together for comfort, but to whom, to what would it cry out? Whereas we fly towards comrades and a kind of celebration. A lamplight gleaming from the humblest hut can change the rudest winter night into Christmas Eve. We in this plane are bound for a place where there will be comrades to welcome us. We in this plane are bound for the communion of our daily bread.

Sufficient unto this day is the weariness and the bliss thereof. I shall turn over to the ground crew my ship made noble by her scars. I shall strip off my cumbrous flying clothes; and as it is now too late to win that drink from Pénicot, I shall go directly to table and dine among my comrades. We are late. Those who are late never get back. Late, are they? If late, then too late. Then nothing can be done for them. The night has swung them into eternity.

Yet at the dinner hour, when the Group takes a census of its dead, one thing is done for them: they are made handsomer than was their wont. They are sketched for ever in their most luminous smile. But we in this plane are surrendering that privilege. We shall surge up out of nowhere, like demons, like poachers in a wood. The major's hand will stop with his bread half way to his mouth. He will stare at us. Perhaps he will say, "Oh! . . . Oh, there you are!" The rest will say nothing. They will scarcely throw us a glance.

There was a time when I had small respect for grown-ups. I was wrong. Men do not really grow old. Men are as pure when you come back to them as when you left them. "Oh, there you are, you who are of our kind!" The words thought and not spoken, out of delicacy of feeling.

Major Alias, that communion of spirit with the Group was to me as is the fire in the hearth to the blind. The blind sit down and put forth their palms, not seeing the source of the gladness they feel. We come home from our sortie ready for our silent reward. Its quality is unique, for it is the quality of love. We do not recognize it as love. Love, when ordinarily we think of it, implies a more tumultuous pathos. But this is the veritable love—a web woven of strands in which we are fulfilled.

## XXII

When I got back to my billet I found my farmer at table with his wife and niece.

"Tell me," I said to him; "how many instruments do you think a pilot has to look after?"

"How should I know? Not my trade," he answered. "Must be some missing, though, to my way of thinking. The ones you win a war with. Have some supper?"

I said I'd had supper at the mess, but already he wasn't listening to me.

"You, our niece, there. Shove along a little. Make room for the captain."

I was made to sit down between the girl and her aunt. Here was something besides the Group that I formed part of. Through my comrades I was woven into the whole of my country. Love is a seed: it has only to sprout, and its roots spread far and wide.

Silently my farmer broke the bread and handed it round. Unruffled, austere, the cares of his day had clothed him in dignity. Perhaps for the last time at this table, he shared his bread with us as in an act of worship. I sat thinking of the wide fields out of which that substance had come. To-morrow those fields would be invaded by the enemy. Oh, there would be no tumult of men and clashing arms! The earth is vast. My farmer would see no more of the invasion than a solitary sentinel posted against the wide sky on the edge of the fields. In appearance nothing would have changed; but a single sign is enough to tell man that everything has changed.

The wind running through the field of grain will still resemble a wind running over the sea. But the wind in the grain is a more wonderful sweep, for as it ruffles the tips of the wheat it takes a census of a patrimony. It takes stock of a future. The wind in the grain is the caress to the spouse, it is the hand of peace stroking her hair.

To-morrow that wheat will have changed. Wheat is something more than carnal fodder. To nourish man is not the same as to fatten cattle. Bread has more than one meaning. We have learnt to see in bread a means of communion between men, for men break bread together. We have learnt to see in bread the symbol of the dignity of labor, for bread is earned in the sweat of the brow. We have learnt to see in bread the essential vessel of compassion, for it is bread that is distributed to the miserable. There is no savor like that of bread shared between men. And I saw of a sudden that the energy contained in this spiritual food, this bread of the spirit generated by that field of wheat, was in peril. To-morrow, perhaps, when he broke bread again and sent it round the table, my farmer would not be celebrating the same household rite. To-morrow, perhaps, his bread would not bring the same glow into these faces round the table. For bread is like the oil of the lamp: its merit is in the light it sheds.

I looked at the beautiful niece beside me and said to myself, "Bread, in this child, is transmuted into languid grace. It is transmuted into modesty. It is transmuted into gentle silence. And to-morrow, perhaps, this same bread, by virtue of a single gray blot rising on the edge of that ocean of wheat, though it nourish this same lamp, will perhaps no longer send forth this same glowing light. The power that is in this bread will have gone out of it."

I had made war this day to preserve the glowing light in that lamp, and not to feed that body. I had made war for the particular radiation into which bread is transmuted in the homes of my countrymen. What moved me so deeply in that pensive little girl was the insubstantial vestment of the spirit. It was the mysterious totality composed by the features of her face. It was the poem on the page, more than the page itself.

The little girl felt that I was looking at her. She raised her eyes to mine. It seemed to me that she smiled at me. Her smile was hardly more than a breath over the face of the waters; but that fugitive gleam was enough. I was moved. I felt, mysteriously present, a soul that belonged in this place and other. There was a peace here, sensing which I murmured to myself, "The peace of the kingdom of silence." That smile was the glow of the shining wheat.

The face of the niece was unruffled again, veiling its unfathomable depth. The farmer's wife sighed, looked round at us, and spoke no word. The farmer, his mind on the day to come, sat wrapped in his earthy wisdom. Behind the silence of these three beings there was an inner abundance that was like the patrimony of a whole village asleep in the night—and like it, threatened. Strange, the intensity with which I felt myself responsible for that invisible patrimony. I went out of the house to walk alone on the highway, and I carried with me a burden that seemed to me tender and in no wise heavy, like a child asleep in my arms.

I walked slowly, not caring where I went. I had promised myself this conversation with my village; but now I found that I had nothing to say. I was like that heavily-laden bough that had flashed into my mind when the sense of victory had swelled in me. I strolled and lingered, filled with the thought of the ties that bound me to my people. I was one with them, they were one with me. That farmer handing round the bread had made no gift to us at table: he had shared with us and exchanged with us that bread in which all of us had our part. And by that sharing the farmer had not been impoverished but enriched. He had eaten sweeter bread, bread of the community, by that sharing. And I, when I took off for France this afternoon, had made no gift either. We of the Group gave nothing to our people. We were their part in the sacrifice of war. Seeing this, I could see why Hochedé fought the war without mouth-filling words, flew his sorties like a blacksmith working at his smithy. "Who are you?"—"I am the village blacksmith." The blacksmith is serene.

I strolled and lingered on the highway, filled with hope among those who seemed to be hopeless; yet even in this I was not cut off from the rest. I was their part in hope. True, we were already beaten. True, all was in suspense. True, all was threatened. Yet despite this, I could not but feel in myself the serenity of victory. Contradiction in terms? I don't give a fig for terms. I was like Pénicot, Hochedé, Alias, Gavoille. Like them, I had no language by which to justify my feeling of victory. But like them I was filled with the sense of my responsibility. And what man can feel himself at one and the same time responsible and hopeless?

Defeat. . . . Victory. . . . Terms I do not know what to make of. One victory exalts, another corrupts. One defeat kills, another brings life. Tell me what seed is lodged in your victory or your defeat, and I will tell you its future. Life is not definable by situations but by mutations. There is but one victory that I know is sure, and that is the victory that is lodged in the energy of the seed. Sow the seed in the wide black earth and already the seed is victorious, though time must contribute to the triumph of the wheat.

This morning France was a shattered army and a chaotic population. But if in a chaotic population there is a single consciousness animated by a sense of responsibility, the chaos vanishes. A rock pile ceases to be a rock pile the moment a single man contemplates it, bearing within him the image of a cathedral. I shall not fret about the loam if somewhere in it a seed lies buried. The seed will drain the loam and the wheat will blaze.

He who accedes to contemplation transmutes himself into seed. He who make a discovery pulls me by the sleeve to draw my attention to it. He who invents preaches his invention. How a Hochedé will express himself or act, I do not know, nor does it matter. He will surely spread his tranquil faith. What I do see more clearly now is the prime agent of victory. He who bears in his heart a cathedral to be built is already victorious. He who seeks to become sexton of a finished cathedral is already defeated. Victory is the fruit of love. Only love can say what face shall emerge from the clay. Only love can guide man towards that face. Intelligence is valid only as it serves love.

The sculptor is great with the burden of his creation. It matters little that he know not how he will draw it forth from the clay. From one thumb stroke to the next, from error to error, contradiction to contradiction, he will move through the clay towards his creation. Intelligence is not creative; judgment is not creative. If the sculptor be but skill and mind, his hands will be without genius.

Concerning the part played by intelligence, we were long in error. We neglected the substance of man. We believed that the virtuosity of base natures could aid in the triumph of noble causes, that shrewd selfishness could exalt spirits to sacrifice, that withered hearts could by a wind of phrases found brotherhood or love. We neglected Being. The seed of the cedar will become cedar. The seed of the bramble can only become bramble. I shall no longer content myself with judging men according to the phrases by which they justify their acts. I shall no longer accept as gold the bond they put up in the form of words, nor be deceived concerning the direction in which their acts tend. Here is a man striding towards his home: I cannot say if he is going towards quarrel or towards love. I can ask myself only this: "What sort of man is he?" And when I know that, only then shall I know by what lodestone he is impelled, and where he is bound. For in the end man always gravitates in the direction commanded by the lodestone within him.

The seed haunted by the sun never fails to find its way between the stones in the ground. And the pure logician, if no sun draws him forth, remains entangled in his logic. I shall not forget the lesson taught me by my enemy himself. What direction should the armored column take to invest the rear of the enemy? Nobody can say. What should the armored column be for this purpose? It should be weight of sea pressing against dike.

What ought we do? This. That. The contrary of this or that. There is no determinism that governs the future. What ought we be? That is the essential question, the question that concerns spirit and not intelligence. For spirit impregnates intelligence with the creation that is to come forth. And later, intelligence is brought to bed of creation. How should man go about building the first ship ever known? Very complicated, this. The ship will be born of a thousand errors and fumblings. But what should man be to build that first ship? Here I seize the problem of creation at the root. Merchant. Soldier. In love with the prospect of faraway lands. For then of necessity designers and builders will be born of that love. They will drain the energy of workmen and one day launch a ship. What should we do to annihilate a forest? The question is not easy. What be? Obviously, a forest fire.

To-morrow we of France will enter into the night of defeat. May my country still exist when day dawns again. What ought we do to save my country? I do not know. Contradictory things. Our spiritual heritage must be preserved, else our people will be deprived of their genius. Our people must be preserved else our heritage will become lost. For want of a way to reconcile heritage and people in their formulas, logicians will be tempted to sacrifice either the body or the soul. But I want nothing to do with logicians. I want my country to exist both in the flesh and in the spirit when day dawns. Therefore I must bear with all the weight of my love in that direction. There is no passage the sea cannot clear for itself if it bear with all its weight.

The blind move towards the fire in the hearth because the need of that fire is in them. At a distance, they are already governed by it. They seek it because already they have found it. The sculptor guided by the need to mould the clay is already in possession of his creation. And we of the Group are like that. We are warmed by the awareness of the ties that bind us to our people—wherefore we feel ourselves already victorious. We know that we are one with the rest. But that the rest may know it, we must learn to express it. That is a matter of consciousness and language. A matter also of avoiding the verbal traps of superficial logic and polemical wrangling in which substance is destroyed. Above all we must not reject any part of that to which we belong.

And therefore I, leaning back against a wall in the silence of the village night, home from my flight to Arras, enlightened, as it seemed to me, by my flight to Arras, imposed upon myself these rules that I shall never betray.

Since I am one with the people of France, I shall never reject my people, whatever they may do. I shall never preach against them in the hearing of others. Whenever it is possible to take their defence, I shall defend them. If they cover me with shame I shall lock up that shame in my heart and be silent. Whatever at such a time I shall think of them, I shall never bear witness against them. Does a husband go from house to house crying out to his neighbors that his wife is a strumpet? Is it thus that he can preserve his honor? No, for his wife is one with his home. No, for he cannot establish his dignity against her. Let him go home to her, and there unburden himself of his anger.

Thus, I shall not divorce myself from a defeat which surely will often humiliate me. I am part of France, and France is part of me. France brought forth men called Pascal, Renoir, Pasteur, Guillaumet, Hochedé. She brought forth also men who were inept, were politicasters, were cheats. But it would be too easy for a man to declare himself part of the first France and not of the other

Defeat divides men. Defeat unbinds that which was bound. In this unbinding there is danger of death. I shall not contribute to these divisions between Frenchmen by casting the responsibility for the disaster upon those of my people who think differently from me. Where there is no judge, nothing is to be gained by hurling accusations. All Frenchmen were defeated together. I was defeated. Hochedé was defeated. Hochedé does not blame others for the defeat. Hochedé says to himself, "I Hochedé, who am one with France, was weak. France that is one with me, Hochedé, was weak. I was weak in her, and she in me." Hochedé knows perfectly that once he begins distinguishing between his people and himself, he glorifies only himself. And from that moment there ceases to exist a Hochedé who is part of a home, a family, a Group, a nation: there remains a Hochedé who is part of a desert.

If I take upon myself a share in my family's humiliation I shall be able to influence my family. It is part of me, as I am of it. But if I reject its humiliation, my family must collapse; and I shall wander alone, filled with vainglory, but a shell as empty as a corpse.

I reject non-being. My purpose is to be. And if I am to be, I must begin by assuming responsibility. Only a few hours ago I was blind. I was bitter. But now I am able to judge more clearly. Just as I refuse to complain of other Frenchmen, since now I feel myself one with France, so I am no longer able to conceive that France has the right to complain of the rest of the world. Each is responsible for all. France was responsible for all the world. Had France been France, she might have stood to the world as the common ideal round which the world would have rallied. She might have served as the keystone in the world's arch. Had France possessed the flavor of France, the radiation of France, the whole world would have been magnetized into a resistance of which the spearhead would have been France. I reject henceforth my reproaches against the world. Assuming that at a given moment the world lacked a soul, France owed it to herself to serve as the world's soul.

France, too, had need to avoid non-being, and to be. There was a time when my Group volunteered for service elsewhere against aggression—in Norway, and again in Finland. What were Norway and Finland, I used to wonder, to the soldiers and petty officers of France? And I would say to myself that in some confused way those men were volunteering to die in a human cause symbolized by mental images of snow and Christmas sleigh-bells. The salvaging of that particular flavor in the world seemed to justify, in their eyes, the sacrifice of their lives. Had we of France meant a kind of Christmas to the world, the world would have been saved through our being.

The spiritual communion of men the world over did not operate in our favor. But had we stood for that communion of men, we should have saved the world and ourselves. In that task we failed. Each is responsible for all. Each is by himself responsible. Each by himself is responsible for all. I understand now for the first time the mystery of the religion whence was born the civilization I claim as my own: "To bear the sins of man." Each man bears the sins of all men.

## XXIII

Who would call this a creed for the weak? A chief is a man who assumes responsibility. He says, "I was beaten." He does not say, "My men were beaten." Thus speaks a real man. Hochedé would say, "I was responsible."

I know the meaning of humility. It is not self-disparagement. It is the motive power of action. If, intending to absolve myself,

I plead fate as the excuse for my misfortunes, I subject myself to fate. If I plead treason as their excuse, I subject myself to treason. But if I accept responsibility, I affirm my strength as a man. I am able to influence that of which I form part. I declare myself a constituent part of the community of mankind.

Thus there is a creature within me against whom I struggle in order that I may rise superior to myself. Except for that flight to Arras I should never have been able to distinguish between that creature and the man I seek to be. A metaphor comes into my mind. What it is worth, I do not know, but here it is: the individual is a mere path. What matters is Man, who takes that path.

The kind of truth advanced in verbal bickerings can no longer satisfy me. I know now that the freezing of my controls is not to be explained by the negligence of government clerks, nor the absence of friendly nations at the side of France by the egoism of those nations. It is true that we can explain defeat by pointing to the incapacity of specific individuals. But a civilization is a thing that kneads and moulds men. If the civilization to which I belong was brought low by the incapacity of individuals, then my question must be, why did my civilization not create a different type of individual?

A civilization, like a religion, accuses itself when it complains of the tepid faith of its members. Its duty is to indue them with fervor. It accuses itself when it complains of the hatred of other men not its members. Its duty is to convert those other men. Yet there was a time when my civilization proved its worth—when it inflamed its apostles, cast down the cruel, freed peoples enslaved—though to-day it can neither exalt nor convert. If what I seek is to dig down to the root of the many causes of my defeat; if my ambition is to be born anew, I must begin by recovering the animating power of my civilization, which has become lost.

For what is true of wheat is true also of a civilization. Wheat nourishes man, but man in turn preserves wheat from extinction by storing up its seed. The seed stored up is a kind of heritage received by one generation of wheat after another. If wheat is to flourish in my fields, it is not enough that I be able to describe it and desire it. I must possess the seed whence it springs. And so with my civilization, for it too springs from energy contained within a seed. If what I wish is to preserve on earth a given type of man, and the particular energy that radiates from him, I must begin by salvaging the principles that animate that kind of man.

My civilization had ceased to be radiant energy. I was able to describe it glibly enough; but I had lost sight of the principle that animated it and bore it along through the ages. And what I have learnt this night is that the words I used to describe my civilization never went to the heart of the matter. Thus I have preached Democracy, for example, without the least notion that, in respect of the qualities and destiny of Man, I was merely giving expression to an aggregate of wishes and not an aggregate of principles. I wished man to be fraternal, free, and strong. Of course! Who would not wish the same? I was able to describe how man ought to act—but not what he ought to be. I used words like mankind, but without defining them. The idea of a community of men seemed to me natural and self-evident. But what is there natural and self-evident about it? The moral climate I had in mind is not natural—it is the product of a particular architecture. A fascist band, a slave market is a community of men—of a sort.

As for my community of men, I waited until I was in jeopardy before I took thought of it. As soon as danger threatened, I took shelter behind it. "What!" I cried. "Are you not ashamed to attack such a beautiful cathedral!" But I had long ceased to be the architect of that cathedral. I had been living in it as sexton, as beadle. Which is to say, as a man defeated in advance. I had been taking advantage of its tranquillity, its tolerance, its warmth. I had been a parasite upon it. It had meant to me no more than a place where I was snug and secure, like a passenger on a ship. The passenger makes use of the ship and gives it nothing in return. The ship is to him a water-tight playground. He is indifferent to the straining of the timbers against the ceaseless hostility of the sea. How he would cry out if the ship were capsized by a storm! But what has he sacrified to the ship? If the members of my civilization have degenerated, and if I have been defeated, against whom am I to lodge a complaint?

There exists a common denominator that integrates all the qualities I demand in the men of my civilization. There exists a keystone that sustains the arch of the particular community which men are called to found. There exists a principle, an animating force, out of which everything once emerged—root, trunk, branches, fruit. That principle was once a radiating seed in the loam of mankind. Only by it can I be made victorious. What is it?

It seemed to me that I was learning many thing in the course of my strange village night. There was something extraordinary in the quality of its silence. The least sound filled all space like a bell. Nothing existed that was not part of me—neither the moaning of the cattle, nor a sudden distant cry, nor the sound of a door as it shut. Each little happening seemed to happen within me, and each stirred up a feeling so poignant that I sought to seize it and fix it before it could vanish.

"That gun-fire over Arras," I said to myself. It had cracked my stubborn shell, and I was released. Within that shell, I must have been setting my house in order the whole day through. I had been the grumbling agent of an absentee landlord. I had been, in other words, an individual. And then Man had appeared. Very simply, he had taken the place of the individual within me. He had sent one look down upon that mob on the highway, and had seen in that mob a people. His people. Man, the common denominator uniting me with that people. Because Man inhabited me I had flown homeward to the Group with the feeling that I was hurrying to a fire in a hearth. Because Man was looking at men through my eyes—Man, the common denominator of all comrades.

Was it a sign? I was so ready to believe in signs. The night was filled with an apprehension of tacit concord. Each sound reached me like a message at once limpid and obscure. I heard suddenly the footsteps of a man on his way home.

"Good evening, Captain."

"Good evening."

I did not know the man. We were like two fishermen hailing each other from bark to bark. Yet once again I sensed the existence of a miraculous relationship. Man, dwelling this night within me, would never make an end of counting his own. Man, the common denominator of peoples and nations.

That man was on his way home with his budget of cares and ruminations and images. With his own cargo locked up within himself. I might have gone up to him and spoken. On the white strip of a village street we might have exchanged a few of our memories. So merchants on the way home from faraway lands used to exchange treasures when they met.

In my civilization, he who is different from me does not impoverish me—he enriches me. Our unity is constituted in something higher than ourselves—in Man. When we of Group 2-33 argue of an evening, our arguments do not strain our fraternity, they reënforce it. For no man seeks to hear his own echo, or to find his reflection in the glass. Staring into the glass called Man, the

Frenchman of France sees the Norwegian of Norway; for Man heightens and absorbs them both, finds room in himself for the customs of the French as easily as for the manners of the Norwegians. Tales of snow are told in Norway, tulips are grown in Holland, flamencos are sung in Spain—and we, participating in Man, are enriched by them all. This, perhaps, was why my Group longed and volunteered to fight for Norway.

And now I seem to have come to the end of a long pilgrimage. I have made no discovery. Like a man waking out of sleep, I am once again looking at that to which I had for so long been blind. I see now that in my civilization it is Man who holds the power to bind into unity all the individual diversities. There is in Man, as in all beings, something more than the mere sum of the materials that went to his making. A cathedral is a good deal more than the sum of its stones. It is geometry and architecture. The cathedral is not to be defined by its stones, since those stones have no meaning apart from the cathedral, receive from it their sole significance. And how diverse the stones that have entered into this unity! The most grimacing of the gargoyles are easily absorbed into the canticle of the cathedral.

But the significance of Man, in whom my civilization is summed up, is not self-evident: it is a thing to be taught. There is in mankind no natural predisposition to acknowledge the existence of Man, for Man is not made evident by the mere existence of men. It is because Man exists that we are men, not the other way round. My civilization is founded upon the reverence for Man present in all men, in each individual. My civilization has sought through the ages to reveal Man to men, as it might have taught us to perceive the cathedral in a mere heap of stones. This has been the text of its sermon—that Man is higher than the individual.

And this, the true significance of my civilization, is what I had little by little forgotten. I had thought that it stood for a sum of men as stone stands for a sum of stones. I had mistaken the sum of stones for the cathedral, wherefore little by little my heritage, my civilization, had vanished. It is Man who must be restored to his place among men. It is Man that is the essence of our culture. Man, the keystone in the arch of the community. Man, the seed whence springs our victory.

It is easy to establish a society upon the foundation of rigid rules. It is easy to shape the kind of man who submits blindly and without protest to a master, to the precepts of a Koran. The real task is to succeed in setting man free by making him master of himself.

But what do we mean by setting man free? You cannot free a man who dwells in a desert and is an unfeeling brute. There is no liberty except the liberty of some one making his way towards something. Such a man can be set free if you will teach him the meaning of thirst, and how to trace a path to a well. Only then will he embark upon a course of action that will not be without significance. You could not liberate a stone if there were no law of gravity—for where will the stone go, once it is quarried?

My civilization sought to found human relations upon the belief in Man above and beyond the individual, in order that the attitude of each person towards himself and towards others should not be one of blind conformity to the habits of the ant-hill, but the free expression of love. The invisible path of gravity liberates the stone. The invisible slope of love liberates man. My civilization sought to make every man the ambassador of their common prince. It looked upon the individual as the path or the message of a thing greater than himself. It pointed the human compass towards magnetized directions in which man would ascend to attain his freedom.

I know how this field of energy came to be. For centuries my civilization contemplated God in the person of man. Man was created in the image of God. God was revered in Man. Men were brothers in God. It was this reflection of God that conferred an inalienable dignity upon every man. The duties of each towards himself and towards his kind were evident from the fact of the relations between God and man. My civilization was the inheritor of Christian values.

It was the contemplation of God that created men who were equal, for it was in God that they were equal This equality possessed an unmistakable significance. For we cannot be equal except we be equal *in* something. The private and the captain are equal in the Nation. Equality is a word devoid of meaning if nothing exists in which it can be expressed.

This equality in the rights of God—rights that are inherent in the individual—forbade the putting of obstacles in the way of the ascension of the individual; and I understand why. God had chosen to adopt the individual as His path. But as this choice also implied the equality of the rights of God "over" the individual, it was clear that individuals were themselves subjected to common duties and to a common respect for law. As the manifestation of God, they were equal in their rights. As the servants of God, they were also equal in their duties.

I understand why an equality that was founded upon God involved neither contradiction nor disorder. Demagogy enters at the moment when, for want of a common denominator, the principle of equality degenerates into a principle of identity. At that moment the private refuses to salute the captain, for by saluting the captain he is no longer doing honor to the Nation, but to the individual.

As the inheritor of God, my civilization made men equal in Man.

I understand the origin of the respect of men for one another. The scientist owed respect to the stoker, for what he respected in the stoker was God; and the stoker, no less than the scientist, was an ambassador of God. However great one man may be, however insignificant another, no man may claim the power to enslave another. One does not humble an ambassador. And yet this respect for man involved no degrading prostration before the insignificance of the individual, before brutishness or ignorance—since what was honored was not the individual himself but his status as ambassador of God. Thus the love of God founded relations of dignity between men, relations between ambassadors and not between mere individuals.

As the inheritor of God, my civilization founded the respect for Man present in every individual.

I understand the origin of brotherhood among men. Men were brothers in God. One can be a brother only *in* something. Where there is no tie that binds men, men are not united but merely lined up. One cannot be a brother to nobody. The pilots of Group 2-33 are brothers in the Group. Frenchmen are brothers in France.

As the inheritor of God, my civilization made men to be brothers in Man.

I understand the meaning of the duties of charity which were preached to me. Charity was the service of God performed through the individual. It was a thing owed to God, however insignificant the individual who was its recipient. Charity never humiliated him who profited from it, nor ever bound him by the chains of gratitude, since it was not to him but to God that the gift was made. And the practice of charity, meanwhile, was never at any time a kind of homage rendered to insignificance, to brutishness, or to ignorance. The physician owed it to himself to risk his life in the care of a plague-infested nobody. He was serving God thereby. He was never a lesser man for having spent a sleepless night at the bedside of a thief.

As the inheritor of God, my civilization made charity to be a gift to Man present in the individual.

I understand the profound meaning of the humility exacted from the individual. Humility did not cast down the individual, it raised him up. It made clear to him his rôle as ambassador. As it obliged him to respect the presence of God in others, so it obliged him to respect the presence of God in himself, to make himself the messenger of God or the path taken by God. It forced him to forget himself in order that he might wax and grow; for if the individual exults in his own importance, the path is transformed into a sea.

As the inheritor of God, my civilization preached self-respect, which is to say respect for Man present in oneself.

I understand, finally, why the love of God created men responsible for one another and gave them hope as a virtue. Since it made of each of them the ambassador of the same God, in the hands of each rested the salvation of all. No man had the right to despair, since each was the messenger of a thing greater than himself. Despair was the rejection of God within oneself. The duty of hope was translatable thus: "And dost thou think thyself important? But thy despair is self-conceit!"

As the inheritor of God, my civilization made each responsible for all, and all responsible for each. The individual was to sacrifice himself in order that by his sacrifice the community be saved; but this was no matter of idiotic arithmetic. It was a matter of the respect for Man present in the individual. What made my civilization grand was that a hundred miners were called upon to risk their lives in the rescue of a single miner entombed. And what they rescued in rescuing that miner was Man.

I understand by this bright light the meaning of liberty. It is liberty to grow as the tree grows in the field of energy of its seed. It is the climate permitting the ascension of Man. It is like a favorable wind. Only by the grace of the wind is the bark free on the waters.

A man built in this wise disposes of the power of the tree. What space may his roots not cover! What human pulp may he not absorb to grow and blossom in the sun!

But I had ruined everything. I had dissipated the inheritance. I had allowed the notion of Man to rot.

And yet my civilization had expended a good share of its genius and its energy to preserve the cult of a Prince revealed in the existence of individual men, and the high quality of human relations established by that cult. All the efforts of Humanism tended towards this end in the age of the Renaissance and after. Humanism assigned to itself the exclusive mission of brightening and perpetuating the ideal of the primacy of Man over the individual. What Humanism preached was Man.

But as soon as we seek to speak of Man, our language displays itself insufficient. Man is not the same as men. We say nothing essential about the cathedral when we speak of its stones. We say nothing essential about Man when we seek to define him by the qualities of men. Humanism strove in a direction blocked in advance when it sought to seize the notion of Man in terms of logic and ethics, and by these terms communicate that notion to the human consciousness. Unity of being is not communicable in words. If I knew men to whom the notion of the love of country or of home was strange, and I sought to teach them the meaning of these words, I could not summon a single argument that would waken the sense of country or home in them. I may, if I like, speak of a farm by referring to its fields, its streams, its pastures, its cattle. Each of these by itself, and all of them together, contribute to the existence of the farm. Yet in that farm there must be something which escapes material analysis, since there are farmers who are ready to ruin themselves for their farms. And it is that "something else" which is the essence of the farm and enhances the particles of which the farm is composed. The cattle, by that something else, become the cattle of a farm, the meadows the meadows of a farm, the fields the fields of a farm.

Thus man becomes the man of a country, of a group, of a craft, of a civilization, of a religion. But if we are to clothe ourselves in these higher beings we must begin by creating them within ourselves. The being of which we claim to form part is created within us not by words but only by acts. A being is not subject to the empire of language, but only to the empire of acts. Our Humanism neglected acts. Therefore it failed in its attempt.

The essential act possesses a name. Its name is sacrifice.

Sacrifice signifies neither amputation nor repentance. It is in essence an act. It is the gift of oneself to the being of which one forms part. Only he can understand what a farm is, what a country is, who shall have sacrificed part of himself to his farm or country, fought to save it, struggled to make it beautiful. Only then will the love of farm or country fill his heart. A country—or a farm—is not the sum of its parts. It is the sum of its gifts.

So long as my civilization leant upon God it was able to preserve the notion of sacrifice whereby God is created in the hearts of men. Humanism neglected the essential rôle of sacrifice. It thought itself able to communicate the notion of Man by words and not by acts. In order to save the vision of Man present in all men, it could do no more than capitalize the word. And mankind was meanwhile moving down a dangerous slope—for we were in danger of mistaking the average of mankind or the arithmetical sum of mankind for Man. We were in danger of mistaking the sum of the stones for the cathedral. Wherefore little by little we lost our heritage.

Instead of affirming the rights of Man present in the individual we had begun to talk about the rights of the collectivity. We had bit by bit introduced a code for the collectivity which neglected the existence of Man. That code explains clearly why the individual should sacrifice himself for the community. It does not explain clearly and without ambiguity why the community should sacrifice itself for a single member. Why it is equitable that a thousand die to deliver a single man from unjust imprisonment. We still remember vaguely that this should be, but progressively we forget it more and more. And yet it is this principle alone which differentiates us from the ant-hill and which is the source of the grandeur of mankind. For want of an effective concept of humanity—which can rest only upon Man—we have been slipping gradually towards the ant-hill, whose definition is the mere sum of the individuals it contains.

What did we possess that we could set up against the religions of the State and of the Party? What had become of our great ideal of Man born of God? That ideal is scarcely recognizable now beneath the vocabulary of windy words that covers it.

Little by little forgetting man, we limited our code to the problems of the individual. We have gone on preaching the equality of men. But having forgotten Man, we no longer knew what it was we were preaching. Having forgotten in what men were equal, we enunciated a vague affirmation that was of no use to us. How can there be any material equality between individuals as such—the sage and the brute, the imbecile and the genius? On the material plane, equality implies that all men are identical and occupy the same place in the community; which is absurd. Wherefore the principle of equality degenerates and becomes the principle of identity.

We have gone on preaching the liberty of men. But having forgotten Man, we have defined our liberty as a sort of vague license limited only at the point where one man does injury to another. This seeming ideal is devoid of meaning, for in fact no man can act without involving other men. If I, being a soldier, mutilate myself, I am shot. An isolated individual does not exist. He who is sad, saddens others.

And even liberty of this sort had to be subjected to a thousand subterfuges before we could make use of it. We found it impossible to say when this right was valid and when it was not valid, and as we wanted very much to preserve the vague principle of the thing from the innumerable assaults which every society necessarily makes upon the liberty of the individual, we turned hypocrite and shut our eyes.

As for charity, we have not even dared go on preaching it. There was a time when the sacrifice which created beings took the name of charity each time that it honored God in His image upon earth. By our charity to the individual we made our gift to God, and later to Man. But having forgotten both God and Man, we found ourselves giving only to the individual. And from that moment charity became an unacceptable course. It is society and not the mood of the individual that should ensure equity in the sharing of the goods of this world. The dignity of the individual demands that he be not reduced to vassalage by the largesse of others. What a paradox—that men who possessed wealth should claim the right, over and above their possessions, to the gratitude of those who were without possessions!

But above all our miscomprehended charity turned against its own goal. It was founded exclusively upon feelings of pity with regard to individuals—wherefore it forbade us all educative chastisement. But true charity, being the practice of the rites rendered to Man over and above the individual, taught that the individual must be fought in order that Man grow great.

And thus Man became lost to us. And losing Man we emptied all warmth out of that very fraternity which our civilization had preached to us—since we are brothers *in* something, and not brothers in isolation. It is not by contributions to a pool that fraternity is ensured. Fraternity is the creation of sacrifice alone. It is the creation of the gift made to a thing greater than ourselves. But we, mistaking the very root of all true existence, seeing in it a sterile diminution of our goods, reduced our fraternity to no more than a mutual tolerance of one another.

We ceased to give. Obviously, if I insist upon giving only to myself, I shall receive nothing. I shall be building nothing of which I am to form part, and therefore I shall be nothing. And when, afterwards, you come to me and ask me to die for certain interests, I shall refuse to die. My own interest will command me to live. Where will I find that rush of love that will compensate my death? Men die for a home, not for walls and tables. Men die for a cathedral, not for stones. Men die for a people, not for a mob. Men die for love of Man—provided that Man is the keystone in the arch of their community. Men die only for that by which they live.

The sole reason why our society still seemed a fortunate one, and man seemed still to be distinguishable from the collectivity, was that our true civilization, which we were betraying in our ignorance, still sent forth its dying rays and still, despite ourselves, continued to preserve us.

How was it possible for our enemies to understand this when we ourselves no longer understood it? All that they could see in us was rocks strewn in a field. They sought in their way to lend meaning to the notion of collectivity—a notion we were no longer able to define because we had forgotten the existence of Man. Some of our enemies went straight and lightheartedly away to the most extreme conclusions of logic. Collectivity to them meant an absolute collection. Each stone was to be identical with every other stone. And each stone was to reign alone over itself. This was anarchy; and the anarchists, quite aware of the reverence due to Man, applied its principles rigorously to the individual. The contradictions that were born of that rigor were even greater than those that exist in our society.

Others collected the strewn stones and heaped them up in a field. They preached the rights of the Mass. The formula cannot satisfy; for if it is intolerable that a single man tyrannize a Mass, it is equally intolerable that the Mass oppress a single man.

Still others gathered together those powerless stones and out of their arithmetical sum they formed a State. And their state, too, fails to transcend the men who compose it, is too the mere expression of a sum. It stands for the power of the collectivity delegated into the hands of an individual. It is the reign of one stone—which claims to be identical with the rest—over a heap of stones. This State preaches a code of collective existence which once again we refuse to accept—but towards which, nevertheless, we are slowly moving for want of remembering Man who alone would justify our refusal.

The faithful of that new religion would object to several miners risking their lives to save a single miner entombed, for in that case the rock pile would be injured. Let one of their wounded seem to be slowing down the advance of their army, and they will finish him off. The good of the community is a thing which they perceive in arithmetic—and it is arithmetic that governs them. They learn by their arithmetic that they would incur loss if they sought to transcend themselves and become greater than they are. Consequently they must hate those who differ from them—since they possess nothing higher than themselves with which to fuse. Every foreign way of life, every foreign race, every foreign system of thought is necessarily an affront to them. They have no power to absorb others, for if we are to convert men to our way we cannot do it by amputating them but must do it by teaching them to express themselves, offering a goal to their aspirations and a territory for the deployment of their energies. To convert is always to set free. A cathedral is able to absorb its stones, which have no meaning but in it. The rock pile absorbs nothing; and for want of power to absorb, it can only crush. It is not astonishing that a rock pile, with its great weight, possesses more power than stones strewn in a field.

And yet it is I who am the stronger.

I am the stronger provided that I am able to find myself. Provided our Humanism restores Man amongst us. Provided we are able to found our community, and, founding it, make use of the sole efficacious instrument—charity. For our community, as it was when our civilization built it, was no mere sum of interests: it was a sum of gifts.

I am the stronger because the tree is stronger than the materials of which it is composed. It drained those materials into itself. It transformed them into itself. The cathedral is more radiant than any heap of stones. I am the stronger because only my civilization possesses the power to bind into its unity all diversity without depriving any element of its individuality.

When I took off for Arras I asked to receive before giving. My demand was in vain. We must give before we can receive, and build before we may inhabit. By my gift of blood over Arras I created the love that I feel for my kind as the mother creates the breast by the gift of her milk. Therein resides the mystery. To create love, we must begin by sacrifice. Afterwards, love will demand further sacrifices and ensure us every victory. But it is we who must take the first step. We must be born before we can exist.

I came back from Arras, having woven my ties with my farmer's family. Through the translucent smile of his niece I saw the wheat of my village. Beyond my village I saw my country, and beyond my country all other countries. I came back to a civilization which had chosen Man as the keystone in its arch. I came back to Group 2-33—that Group that had volunteered to fight for Norway.

I dressed this day for the service of a god to whose being I was blind. Arras unsealed my eyes. Like the others of the Group, I am no longer blind. It may be that to-morrow Alias will order me to fly still another sortie. If, at dawn to-morrow, I fight again, I shall know finally why I fight.

My eyes have been unsealed, and I want now to remember what it is that they have seen. I feel the need of a simple Credo so that I may remember.

I believe in the primacy of Man above the individual and of the universal above the particular.

I believe that the cult of the universal exalts and heightens our particular riches, and founds the sole veritable order, which is the order of life. A tree is an object of order, despite the diversity of its roots and branches.

I believe that the cult of the particular is the cult of death, for it founds its order upon likeness. It mistakes identity of parts for unity of Being. It destroys the cathedral in order to line up the stones. Therefore I shall fight against all those who strive to impose a particular way of life upon other ways of life, a particular people upon other peoples, a particular race upon other races, a particular system of thought upon other systems of thought.

I believe that the primacy of Man founds the only equality and the only liberty that possess significance. I believe in the equality of the rights of Man inherent in every man. I believe that liberty signifies the ascension of Man. Equality is not identity. Liberty is not the exaltation of the individual against Man. I shall fight against all those who seek to subject the liberty of Man either to an individual or to the mass of individuals.

I believe that what my civilization calls charity is the sacrifice granted Man for the purpose of his own fulfillment. Charity is the gift made to Man present in the insignificance of the individual.

It creates Man. I shall fight against all those who, maintaining that my charity pays homage to mediocrity, would destroy Man and thus imprison the individual in an irredeemable mediocrity.

I shall fight for Man. Against Man's enemies—but against myself as well.

# XXIV

We collected again at midnight to receive orders. Group 2-33 was sleepy. The flame in the fireplace had turned to embers. The Group seemed to be holding up still, but this was an illusion. Hochedé was staring glumly at his precious watch. Pénicot stood against a wall in a corner, his eyes shut. Gavoille, sitting on a table, his glance vacant and legs hanging, was pouting like a child about to cry. The doctor was nodding over a book. Alias alone was still alert, but frighteningly pale, papers before him under the lamplight, discussing something in a low voice with Geley. Discussion, indeed, gives you a false picture. The major was talking. Geley was nodding his head and saying, "Yes, of course." Geley was hanging on to that "Yes, of course" by main strength. He was clinging more and more eagerly to the major's discourse, like a half-drowned man to the neck of a swimmer. Had I been Alias I should have said without a change of voice, "Captain Geley, you are to be shot at dawn," and waited for the answer.

The Group had not slept for three nights. It stood like a house of cards.

The major got up, went across to Lacordaire, and pulled him out of a dream in which perhaps he was beating me at chess.

"Lacordaire! You take off at dawn. Ground-scraper sortie."

"Very good, Major."

"Better get some sleep."

"Yes, Major."

Lacordaire sat down again. The major went out, drawing Geley in his wake as if he were a dead fish on the end of a line. It was nearer a week than three days since Geley had been to bed. Like Alias, not only did he fly his sorties, but he carried part of the burden of responsibility for the Group. Human resistance has its limits: Geley seemed to have crossed his. Yet there they were, the swimmer and his burden, going off to the Staff for phantom orders.

Vezin, the skeptical Vezin, asleep on his feet, came teetering over to me like a somnambulist:

"You asleep?"

"I . . ."

I had been lying back in an armchair (for I had found an armchair) and was indeed dropping off. But Vezin's voice bothered me. What was it he had said? "Looks bad, old boy. . . . Categorically blocked. . . . Looks bad. . . ."

"You asleep?"

"I. . . . No. . . . What looks bad?"

"The war," he said.

That was news, now! I started to drop off again and murmured vaguely, "What war?"

"What do you mean, 'What war'!"

This conversation wasn't going to get very far. Ah, Paula! Had air squadrons been issued with Tyrolian nursemaids we should have been put to bed long ago.

The major flung open the door and called out, "All set! We move out to-night!"

Behind him stood Geley, wide-awake. He would put off his "Yes, of course" until to-morrow night. Once again he would somehow find a reserve of strength in himself to help him with the wearying chores of our removal.

The Group got to its feet. The Group said, "Move again? Very good, sir." What else was there to say?

There was nothing to say. We should see to the removal. Lacordaire would stay behind and take off at dawn. If he got back he would meet us at our new base.

There would be nothing to say to-morrow, either. To-morrow, in the eyes of the bystanders, we would be the defeated. The defeated have no right to speak. No more right to speak than has the seed.

The End